# Further praise for *The Metaverse*

"*The Metaverse* provides a clear and informed picture of what the Metaverse will be, and how and why it will emerge. Matthew Ball gives us a deeply insightful analysis of the key technologies that will inevitably bring us the Metaverse, and of how the Metaverse will change lives on a global scale while creating tens of trillions of dollars of value."

—**John Riccitiello,** CEO of Unity Technologies and former CEO of Electronic Arts

"Trying to lasso a definition of the Metaverse is as baffling as a swollen campaign speech. Will the Metaverse replace the internet? Will it fizzle like a bright firecracker? Enter clear-eyed Matthew Ball, a brilliant student of business and the digital age. In sentences that clarify rather than obscure, Ball serves as our sherpa, helping readers decipher both the hype and the potentially seismic changes ahead."

—**Ken Auletta,** *New York Times* best-selling author of *Googled* and *Frenemies*

"Thoughtful, engaging, and relevant. However what we think of as the Metaverse evolves, the issues that Matthew Ball raises in this book will continue to shape our shared future, both online and off."

—**Phil Spencer,** CEO of Microsoft Gaming

# THE
# METAVERSE

## And How It Will
## Revolutionize Everything

# MATTHEW BALL

**LIVERIGHT PUBLISHING CORPORATION**

A DIVISION OF W.W. NORTON & COMPANY

*Independent Publishers Since 1923*

For information about permission to reproduce selections from this book,
write to Permissions, Liveright Publishing Corporation, a division of
W. W. Norton & Company, Inc., 500 Fifth Avenue, New York, NY 10110

For information about special discounts for bulk purchases, please contact
W. W. Norton Special Sales at specialsales@wwnorton.com or 800-233-4830

Manufacturing by Lake Book Manufacturing
Book design by Lovedog Studio
Production manager: Anna Oler

ISBN 978-1-324-09203-2

Liveright Publishing Corporation, 500 Fifth Avenue, New York, N.Y. 10110
www.wwnorton.com

W. W. Norton & Company Ltd., 15 Carlisle Street, London W1D 3BS

1 2 3 4 5 6 7 8 9 0

*To Rosie, Elise, and Hillary*

# CONTENTS

## *Part III*
# HOW THE METAVERSE WILL REVOLUTIONIZE EVERYTHING

# INTRODUCTION

TECHNOLOGY FREQUENTLY PRODUCES SURPRISES that no one predicts. But the biggest and most fantastical developments are often anticipated decades in advance. In the 1930s, Vannevar Bush, then president of the Carnegie Institution of Washington, began work on a hypothetical electromechanical device that would store all books, records, and communications, and mechanically link them together by keyword association, rather than traditional, mostly hierarchical storage models. Despite the enormity of its archive, Bush stressed that this "Memex" (short for "memory extender") could be consulted "with exceeding speed and flexibility."

In the years that followed this early research, Bush became one of the most influential engineers and science administrators in American history. From 1939 to 1941, he was vice chairman and temporarily served as chairman of the National Advisory Committee for Aeronautics, the predecessor agency to NASA. In this position, Bush convinced President Franklin D. Roosevelt to establish what became the Office of Scientific Research and Development (OSRD), a new federal agency that would be run by Bush, who would report directly to the president. The agency was provided nearly unlimited funding, primarily for secret projects that would aid the United States' efforts in World War II.

Only four months after OSRD was founded, President Roosevelt approved the atomic bomb program known as the Manhattan Project, following a meeting with Bush and Vice President Henry A. Wallace.

To manage the program, Roosevelt created a Top Policy Group consisting of himself, Bush, Wallace, Secretary of War Henry L. Stimson, Chief of Staff of the Army General George C. Marshall, and James B. Conant, who headed up a subbranch of OSRD previously run by Bush. In addition, the Uranium Committee (later named the S-1 Executive Committee) would report directly to Bush.

After the war ended in 1945, but two years before he left his role as director of the OSRD, Bush wrote two famous essays. The first, "Science, the Endless Frontier," was addressed to the president and in it, Bush called for an increase in government investments in science and technology, rather than a peacetime reduction, as well as the establishment of the National Science Foundation. The second essay, "As We May Think," appeared in *The Atlantic* and publicly detailed Bush's vision of the Memex.

In the years that followed his essays, Bush stepped back from public office and public view. But soon enough, his various contributions to government, science, and society began to converge. Starting in the 1960s, the US government funded a variety of projects within the Department of Defense, in partnership with a network of external researchers, universities, and other nongovernment institutions that together developed the foundation of the internet. At the same time, Bush's Memex was informing the creation and evolution of "hypertext," one of the underlying concepts of the World Wide Web, which is typically written in the HyperText Markup Language (HTML) and enables users to instantly access a nearly infinite extent of online content by clicking on a given piece of text. Twenty years later, the US federal government established the Internet Engineering Task Force to guide the technical evolution of the Internet Protocol Suite, and with the help of the Department of Defense founded the World Wide Web Consortium, which, among other duties, manages the ongoing development of HTML.

While technological progress typically occurs out of common sight, science fiction often provides the general public with the clearest view of the future. In 1968, fewer than 10% of American households had a color TV, yet the second-highest-grossing film of the year, *2001: A Space Odyssey*, imagined a future in which humanity had compressed these fridge-sized devices into coaster-thin displays and used them idly during breakfast. Anyone watching the film today will instantly liken these devices to iPads. Per usual, the imagined technology, like Bush's Memex, took longer to arrive than was originally anticipated. iPads appeared in stores four and half decades after Stanley Kubrick's groundbreaking film was released, and more than a decade after the futuristic film was set.

By 2021, tablets had become commonplace and spacefaring had begun to feel within reach. Throughout that summer, competing efforts from billionaires Richard Branson, Elon Musk, and Jeff Bezos were under way to bring civilian travel to lower orbit and usher in an era of space elevators and interplanetary colonization. However, it was another decades-old science fiction concept, the Metaverse, that seemed to indicate the future had truly arrived.

In July 2021, Facebook founder and CEO Mark Zuckerberg said: "In this next chapter of our company, I think we will effectively transition from people seeing us as primarily being a social media company to being a metaverse company. And obviously, all of the work that we're doing across the apps that people use today contribute directly to this vision."[1] Shortly thereafter, Zuckerberg publicly announced a division focused on the Metaverse and elevated the head of Facebook Reality Labs—a division that works on miscellaneous futuristic projects including Oculus VR (virtual reality), AR (augmented reality) glasses, and brain-to-machine interfaces—to chief technology officer. In October 2021, Zuckerberg proclaimed that Facebook would be changing its name to Meta Platforms* to reflect its shift to this "Metaverse." To the surprise of many Facebook shareholders, Zuckerberg also said that his investments in the Metaverse would reduce operating income by over

---

* For the sake of clarity, this book refers to Meta Platforms as Facebook. Explaining the Metaverse and its various platforms, while also discussing an early leader in the Metaverse that is called Meta Platforms, would only confuse matters.

$10 billion in 2021, while warning that these investments would grow for several more years.

Zuckerberg's bold pronouncements drew the most attention, but many of his peers and competitors had launched similar initiatives and made similar announcements in the months prior. In May, Microsoft CEO Satya Nadella began to speak of a Microsoft-led "enterprise Metaverse." Likewise, Jensen Huang, CEO and founder of computing and semiconductor giant Nvidia, had told investors that "the economy in the Metaverse . . . [will] be larger than the economy in the physical world"* and that Nvidia's platforms and processors would be at the heart of it.[2] In the fourth quarter of 2020 and first quarter of 2021, the gaming industry had two of its largest-ever initial public offerings (IPOs) in Unity Technologies and Roblox Corporation, both of which wrapped their corporate histories and ambitions in Metaverse-related narratives.

Throughout the remainder of 2021, the term "Metaverse" almost became a punchline as every company and its executives seemed to trip over themselves to mention it as something that would make their company more profitable, their customers happier, and their competitors less threatening. Prior to Roblox's IPO filings in October 2020, the "Metaverse" had appeared only five times in US Securities and Exchange Commission filings.[3] In 2021, the term was mentioned more than 260 times. That same year, Bloomberg, a software company that provides financial data and information to investors, catalogued more than a thousand stories containing the word Metaverse. The prior decade had only seven.

Interest in the Metaverse was not limited to Western nations and corporations. In May 2021, China's largest company, the internet gaming giant Tencent, publicly described its vision of the Metaverse, calling it "Hyper Digital Reality." The following day, South Korea's Ministry of Science and ICT (Information and Communications Technol-

---

* In 2021, global GDP was estimated at roughly $90 trillion–$95 trillion by the International Monetary Fund, United Nations, and World Bank.

ogy) announced "The (South Korean) Metaverse Alliance," spanning over 450 companies including SK Telecom, Woori Bank, and Hyundai Motor. In early August, South Korean gaming giant Krafton, maker of *PlayerUnknown's Battlegrounds* (also known as *PUBG*) completed its IPO, the second largest in the country's history. Krafton's investment bankers made sure to tell would-be investors that the company would also be a global leader in the Metaverse. In the ensuing months, Chinese internet giants Alibaba and ByteDance, the parent company of the global social network TikTok, both began to register various Metaverse trademarks and acquire various VR and 3D-related start-ups. Krafton, meanwhile, committed publicly to launching a "PUBG Metaverse."

The Metaverse captured more than the imagination of techno-capitalists and sci-fi fans. Not long after Tencent publicly unveiled its vision of hyper-digital reality, the Communist Party of China (CCP) began its biggest-ever crackdown of its domestic gaming industry. Among several new policies was a prohibition on minors playing video games Monday through Thursday that also limited their play from 8 p.m. to 9 p.m. on Friday, Saturday, and Sunday nights (in other words, it was impossible for a minor to play a video game for more than three hours per week). In addition, companies such as Tencent would use their facial recognition software and a player's national ID to periodically ensure that these rules were not being skirted by a gamer borrowing an older user's device. Tencent also pledged $15 billion in aid for "sustainable social value," which *Bloomberg* said would be focused on "areas like increasing incomes for the poor, improving medical assistance, promoting rural economic efficiency and subsidizing education programs."[4] Alibaba, China's second-largest company, committed a similar amount only two weeks later. The message from the CCP was clear: look to your countrymen and women, not virtual avatars.

The CCP's concerns about the growing role of gaming content and platforms in public life became more explicit in August, when the state-owned *Security Times* warned its readers that the Metaverse is a "grand and illusionary concept" and "blindly investing [in it] will ultimately come back to bite you."*5 Some commentators interpreted China's various warnings, prohibitions, and taxes as confirmation of the Metaverse's significance. For a communist and centrally planned country ruled by a single party, the potential of a parallel world for collaboration and communication is a threat, regardless of whether it's run by a single corporation or decentralized communities.

Yet China was not alone in its worries. In October, members of the European Parliament began to voice concerns. One particularly important voice was that of Christel Schaldemose, who served as a chief negotiator for the European Union as it worked on its largest-ever overhaul of digital-era regulations (most of which were intended to curb the power of so-called big tech giants such as Facebook, Amazon, and Google). In October, she told the Danish paper *Politiken* that "plans for metaverse are deeply, deeply worrying" and that the union "has to take them into account."6

It's possible that the many Metaverse announcements, critiques, and warnings are just a real-world echo chamber about a virtual fantasy—or more about driving new narratives, product launches, and marketing than anything life-changing. After all, the tech industry has a history of using buzzwords that are hyped for far longer than they ultimately end up lasting in the market, such as 3D televisions, or that prove to be further away than originally promised, such as VR headsets or virtual assistants. But it's rare that the world's largest companies publicly reorient themselves around such ideas at an early stage, thereby setting themselves up to be evaluated by employees, customers, and shareholders on the basis of their success in realizing their most ambitious visions.

The dramatic response to the Metaverse reflects the growing belief that it is the next great computing and networking platform, similar in scope to the transition from the personal computer and fixed-line internet of the 1990s to the era of mobile and cloud computing we live in today. That shift popularized a once-obscure business school

---

* The *Security Times* cited the author of this book when describing the Metaverse.

term—"disruption"—and transformed almost every industry while reshaping modern society and politics. Yet there is a critical difference between that shift and the impending shift to the Metaverse: timing. Most industries and individuals did not foresee the significance of mobile and cloud, and consequently were stuck reacting to changes and fighting off disruption from those who better understood them. Preparations for the Metaverse are happening much earlier, and proactively.

In 2018, I began writing a series of online essays on the Metaverse, then an obscure and fringe concept. In the years since, these essays have been read by millions of people as the Metaverse has transitioned from the world of paperback science fiction to the front page of the *New York Times* and corporate strategy reports around the world.

*The Metaverse: And How It Will Revolutionize Everything* updates, expands, and recasts everything I've previously written on the Metaverse. The book's core purpose is to offer a clear, comprehensive, and authoritative definition of this still inchoate idea. Yet my ambitions are broader: I hope to help you understand what's required to realize the Metaverse, why entire generations will eventually move to and live inside it, and how it will forever alter our daily lives, our work, and how we think. In my view, the collective value of these changes will be in the tens of trillions of dollars.

*Part I*

# WHAT
# IS THE
# METAVERSE?

*Chapter 1*

# A BRIEF HISTORY
# OF THE FUTURE

THE TERM "METAVERSE" WAS COINED BY AUTHOR Neal Stephenson in his 1992 novel *Snow Crash*. For all its influence, Stephenson's book provided no specific definition of the Metaverse, but what he described was a persistent virtual world that reached, interacted with, and affected nearly every part of human existence. It was a place for labor and leisure, for self-actualization as well as physical exhaustion, for art alongside commerce. At any given time, there were roughly 15 million human-controlled avatars on "The Street," which Stephenson called "the Broadway, the Champs Elysees of the Metaverse," but stretched across the entirety of a virtual planet more than two and a half times the size of the earth. As a point of contrast, there were fewer than 15 million total users of the internet in the real world the year Stephenson's novel was published.

While Stephenson's vision was vivid and, to many, inspiring, it was also dystopic. *Snow Crash* is set at some point in the early 21st century, years after a global economic collapse. Most layers of government have been replaced by for-profit "Franchise-Organized Quasi-National Entities" and "burbclaves," a contraction of the term "suburban enclaves." Each burbclave operates as a "city-state with its own constitution, a border, laws, cops, everything"[1] and some even provide "citizenship" purely based on race. The Metaverse offers refuge and opportunity to millions. It was a virtual place where a pizza deliverer in the "real world" could be a talented swordsman with inside access to the hottest clubs. But Stephenson's novel was clear: in *Snow Crash* the Metaverse has made life in the real world worse.

As with Vannevar Bush, Stephenson's influence on modern technology only grows with time, even if he is mostly unknown to the public. Conversations with Stephenson helped inspire Jeff Bezos to found the private aerospace manufacturer and suborbital spaceflight company Blue Origin in 2000, with the author working there part-time until 2006, when he became a senior advisor to the company (a position he still holds). As of 2021, Blue Origin is considered the second most valuable company of its kind, ranked only behind Elon Musk's SpaceX. Two of the three founders of Keyhole, now known as Google Earth, have said their visions were informed by a similar product described in *Snow Crash*, and that they once tried to recruit Stephenson to the company. From 2014 to 2020, Stephenson was also "Chief Futurist" at Magic Leap, a mixed reality company that was also inspired by his work. The company later raised over half a billion dollars from corporations including Google, Alibaba, and AT&T, attaining a peak valuation of $6.7 billion, before struggles to realize its vaulting ambitions resulted in a recapitalization and the departure of its founder.* Stephenson's novels have been cited as the inspiration for various cryptocurrency projects and non-cryptographic efforts to build decentralized computer networks, as well as the production of CGI-based movies which are watched at home but generated live through the motion-captured performance of actors that might be tens of thousands of miles away.

Despite his far-reaching impact, Stephenson has consistently warned against a literal interpretation of his works—especially *Snow Crash*. In 2011, the novelist told the *New York Times* that "I can talk all day long about how wrong I got it"[2] and, when asked about his influence

---

* The company's valuation was ultimately reduced by more than two-thirds, with the company's investors hiring Peggy Johnson, a long-time executive vice president at Qualcomm and Microsoft, to lead as CEO. It is during this time that Stephenson left the company, along with many other full-time employees and other chief officers.

on Silicon Valley by *Vanity Fair* in 2017, he reminded the publication to keep "in mind that [*Snow Crash* was written] pre-Internet as we know it, pre-Worldwide Web, just me making shit up."[3] As a result, we should be wary of reading too much into Stephenson's specific vision. And while he coined the term "Metaverse," he was far from the first to introduce the concept.

In 1935, Stanley G. Weinbaum wrote a short story titled "Pygmalion's Spectacles," about the invention of magical VR-like goggles that produced a "movie that gives one sight and sound . . . you are in the story, you speak to the shadows, and the shadows reply, and instead of being on a screen, the story is all about you, and you are in it."*[4] Ray Bradbury's 1950 short story "The Veldt" imagines a nuclear family in which the parents are supplanted by a virtual reality nursery that the children never want to leave. (The children eventually lock their parents inside the nursery, which then kills them.) Philip K. Dick's 1953 story "The Trouble with Bubbles" is set in an era where humans have explored deep into outer space, but never succeeded in finding life. Yearning to connect with other worlds and life-forms, consumers begin to buy a product called "Worldcraft" through which they can build and "Own [Their] Own World," which are cultivated to the point of producing sentient life and fully realized civilizations (most Worldcraft-owners eventually destroy their worlds in what Dick described as a "neurotic" "orgy of breaking" intended to "assume some god suffering from ennui"). A few years later, Isaac Asimov's novel *The Naked Sun* was published. In it, he described a society where face-to-face interactions ("seeing") and physical contact are considered both wasteful and repugnant, and most work and socializing takes place via remotely projected holograms and 3D televisions.

In 1984, William Gibson popularized the term "cyberspace" in his

---

* Pygmalion is a reference to the mythological Cypriot king Pygmalion. In Ovid's epic poem *Metamorphoses*, Pygmalion carves a sculpture so beautiful and lifelike that he falls in love with and marries her; the goddess Aphrodite transforms her into a living woman.

novel *Neuromancer*, defining it as "A consensual hallucination experienced daily by billions of legitimate operators, in every nation. . . . A graphic representation of data abstracted from the banks of every computer in the human system. Unthinkable complexity. Lines of light ranged in the nonspace of the mind, clusters and constellations of data. Like city lights, receding." Notably, Gibson called the visual abstraction of cyberspace "The Matrix," a term repurposed by Lana and Lilly Wachowski 15 years later for their film of the same name. In the Wachowskis' movie, the Matrix refers to a persistent simulation of the planet earth as it was in 1999, but which all of humanity is unknowingly, indefinitely, and forcibly connected to in the year 2199. The purpose of this simulation is to placate the human race so that it can be used as bioelectric batteries by the sentient, but man-made, machines which conquered the planet in the 22nd century.

## The Program Is More Optimistic than the Pen

Whatever the differences among each specific author's visions, the synthetic worlds of Stephenson, Gibson, the Wachowskis, Dick, Bradbury, and Weinbaum are all presented as dystopias. Yet there is no reason to assume that such an outcome is inevitable, or even likely, for the actual Metaverse. A perfect society tends not to make for much human drama, and human drama is the root of most fiction.

As a point of contrast, we can consider the French philosopher and cultural theorist Jean Baudrillard, who coined the term "hyperreality" in 1981 and whose works are often linked to those of Gibson, and those Gibson influenced.* Baudrillard described hyperreality as a state

---

* When asked about Baudrillard in April 1991, Gibson said, "He's a cool science-fiction writer" (Daniel Fischlin, Veronica Hollinger, Andrew Taylor, William Gibson, and Bruce Sterling, "'The Charisma Leak': A Conversation with William Gibson and Bruce Sterling," *Science Fiction Studies* 19, no. 1 [March 1992], 13). The Wachowskis tried to involve Baudrillard in their film, but he declined and later

in which reality and simulations were so seamlessly integrated that they were indistinguishable. Though many find this idea frightening, Baudrillard argued that what mattered was where individuals would derive more meaning and value—and speculated it would be in the simulated world.[5] The idea of the Metaverse is also inseparable from the ideas of the Memex, but where Bush imagined an infinite series of documents linked together via words, Stephenson and others conceived infinitely interconnected worlds.

More instructive than Stephenson's texts and those which inspired them are the many efforts to build virtual worlds over the past several decades. This history not only shows a multi-decade progression towards the Metaverse, but also reveals more about its nature. These would-be Metaverses have not been centered on subjugation or profiteering, but on collaboration, creativity, and self-expression.

Some observers date the history of "proto-Metaverses" to the 1950s during the rise of mainframe computers, which represented the first time that individuals could share purely digital messages with one another across a network of different devices. Most, however, start in the 1970s with text-based virtual worlds known as Multi-User Dungeons. MUDs were effectively a software-based version of the role-playing game Dungeons & Dragons. Using text-based commands that resembled human languages, players could interact with one another, explore a fictional world populated by non-playable characters and monsters, attain power-ups and knowledge, and eventually retrieve a magical chalice, defeat an evil wizard, or rescue a princess.

---

described the film as a misread of his ideas (Aude Lancelin, "The Matrix Decoded: Le Nouvel Observateur Interview with Jean Baudrillard," *Le Nouvel Observateur* 1, no. 2 [July 2004]). When Morpheus introduces the film's protagonist to the "real world," he tells Neo "As in Baudrillard's vision, your whole life has been spent inside the map, not the territory." (Lana Wachowski and Lilly Wachowski, *The Matrix*, directed by Lana Wachowski and Lilly Wachowski [1999; Burbank, CA: Warner Bros., 1999], DVD.) Recall, too, Tencent's original name for its Metaverse vision: "hyper-digital reality."

The growing popularity of MUDs inspired the creation of Multi-User Shared Hallucinations (or MUSHs) or Multi-User Experiences (MUXs). Unlike MUDs, which asked players to carry out specific roles in the context of a specific and usually fantastical narrative, MUSHs and MUXs enabled participants to collaboratively define the world and its objective. Players might choose to set their MUSH in a courtroom, while taking on roles such as defendant, attorney, plaintiff, judge, and members of the jury. One participant might later decide to transform the relatively mundane proceedings into a hostage situation—which would then be diffused by a poem that was mad-libbed by the other players.

The next great leap came in 1986 with the release of the Commodore 64 online game *Habitat*, which was published by Lucasfilm, the production company founded by *Star Wars* creator George Lucas. *Habitat* was described as "a multi-participant online virtual environment" and, in a reference to Gibson's novel *Neuromancer*, "a cyberspace." Unlike MUDs and MUSHs, the world of *Habitat* was graphical, thereby allowing users to actually see virtual environments and characters, though only via pixeled 2D. It also afforded players far greater control over the in-game environment. "Citizens" of *Habitat* were in charge of the laws and expectations of their virtual world, and had to barter with each other for necessary resources and avoid being robbed or killed for their wares. This challenge led to periods of chaos, after which new rules, regulations, and authorities were established by the player community to maintain order.

Though *Habitat* is not as widely remembered as other 1980s video games, such as *Pac-Man* and *Super Mario Bros.*, it transcended the niche appeal of MUDs and MUSHs, ultimately becoming a commercial hit. The title was also the first game to repurpose the Sanskrit term "avatar," which roughly translates to "the descent of a deity from a heaven," to refer to a user's virtual body. Decades later, this usage has become convention—in no small part because Stephenson reapplied it in *Snow Crash*.

The 1990s saw no major "proto-Metaverse" games, but advances continued. That decade, millions of consumers took part in the first isometric 3D (also known as 2.5D) virtual worlds, which gave the illusion of three-dimensional space, but only allowed users to move across two axes. Not long after, full 3D virtual worlds emerged. A number of games, such as 1994's *Web World* and 1995's *Activeworlds*, also empowered users to collaboratively build a visible virtual space in real time, rather than through asynchronous commands and votes, and introduced a number of graphic/symbol-based tools to make world-building easier. Notably, *Activeworlds* also had the express purpose of building Stephenson's Metaverse, asking players to not just enjoy its virtual worlds, but to invest in expanding and populating it. In 1998, *OnLive! Traveler* launched with spatial voice chat, which allowed users to hear where other players were positioned relative to other participants, and for an avatar's mouth to move in response to the words spoken by the player.[6] The following year, Intrinsic Graphics, a 3D gaming software company, completed the spinoff of Keyhole. While Keyhole did not become broadly popular until the middle of the next decade and after its acquisition by Google, it represented the first time anyone on earth could access a virtual reproduction of the entire planet. In the ensuing 15 years, much of the map was updated to partial 3D and connected to Google's much larger database of mapping products and data, enabling users to also overlay information such as real-time traffic.

It was with the launch of (the aptly named) *Second Life* in 2003 that many, especially those in Silicon Valley, began to contemplate the prospect of a parallel existence that would take place in virtual space. In its first year, *Second Life* attracted over one million regular users, and shortly thereafter, numerous real-world organizations established their own businesses and presences inside the platform. This included for-profit corporations such as Adidas, BBC, and Wells Fargo, as well as nonprofits such as the American Cancer Society and Save the Children and even universities, including Harvard, whose law school offered exclusive courses inside *Second Life*. In 2007, a stock exchange was launched on the platform with the aim of helping *Second Life*–based companies raise capital using the platform's Linden Dollars currency.

Crucially, developer Linden Labs did not intermediate transactions in *Second Life*, nor actively manage what was made or sold. Instead, transactions were made directly between buyers and sellers and based on perceived value and need. Overall, Linden Labs operated more like a government than a game-maker. The company did provide some user-facing services, such as identity management, ownership records, and an in-world legal system. But its focus wasn't on building out the *Second Life* universe directly. Instead, it enabled a thriving economy via ever-improving infrastructure, technical capabilities, and tools that would attract more developers and creators who would then create things for other users to do, places for them to visit, and items for them to buy—attracting more users and therefore more spending, which would in turn attract more investment from developers and creators. To this end, *Second Life* also offered users the ability to import virtual objects and textures made outside the platform. By 2005, just two years after it launched, *Second Life*'s annualized GDP exceeded $30 million. By 2009, it exceeded half a billion dollars, with users cashing out $55 million into real-world currency that year.

For all the success of *Second Life*, it was the rise of virtual world platforms *Minecraft* and *Roblox* that brought its ideas to a mainstream audience in the 2010s. In addition to offering significant technical enhancements compared to their predecessors, *Minecraft* and *Roblox* also focused on children and teenage users, and were therefore far easier to use, rather than just offer greater capabilities. The results have been astounding.

Throughout the 2010s, bands of users collaborated in *Minecraft* to build cities as large as Los Angeles—roughly 500 square miles. One video game streamer, Aztter, constructed a stunning cyberpunk city out of an estimated 370 million *Minecraft* blocks, having worked an average of 16 hours per day for a year.[7] Scale is not the sole achievement of the platform. In 2015, Verizon built a cellphone inside *Minecraft* that could make and receive live video calls to the "real world." As the COVID-19 virus spread across China in February 2020, a community of Chinese *Minecraft* players rapidly re-created the 1.2-million-square-foot hospitals built in Wuhan as a tribute to the "IRL" ("in real life") workers, receiving global press coverage.[8] One month later, Reporters Sans Frontières (also known as Reporters Without Borders) commissioned the construction of a museum within *Minecraft* that was composed of over 12.5 million blocks assembled by 24 virtual builders in 16 different countries over some 250 hours combined. The Uncensored Library, as it was called, allowed users in countries such as Russia, Saudi Arabia, and Egypt to read banned literature, as well as works promoting free speech and detailing the lives of journalists such as Jamal Khashoggi, whose murder was ordered by political leaders in Saudi Arabia.

By the end of 2021, more than 150 million people were using *Minecraft* each month—more than six times as many as in 2014, when Microsoft bought the platform. Despite this, *Minecraft* was far from the size of the new market leader, *Roblox*, which had grown from fewer than 5 million to 225 million monthly users over that same period. According to Roblox Corporation, 75% of children ages 9 to 12 in the United States regularly used the platform in Q2 2020. Combined, the two titles amassed more than 6 billion hours of monthly usage each, which spanned more than 100 million different in-game worlds and had been designed by over 15 million users. The *Roblox* game with the most lifetime plays—*Adopt Me!*—was created by two hobbyist players in 2017 and enabled users to hatch, raise, and trade various pets. By the end of 2021, *Adopt Me!*'s virtual world had been visited more than 30 billion times—more than fifteen times the average number of global tourism visits in 2019. Furthermore, developers on *Roblox*, many of whom are also small teams with fewer than 30 members, have received more than $1 billion in payments from the platform. By the end of 2021, Roblox had become the most valuable gaming company outside of China, worth nearly 50% more than storied gaming giants Activision Blizzard and Nintendo.

Despite the enormous growth in *Minecraft*'s and *Roblox*'s audiences and developer communities, many other platforms began to emerge and grow towards the tail end of the 2010s. In December 2018, for example, the blockbuster video game *Fortnite* launched *Fortnite Creative Mode*, its own riff on *Minecraft*'s and *Roblox*'s world-building. Meanwhile, *Fortnite* was also transforming into a social platform for non-game experiences. In 2020, hip-hop star (and Kardashian family member) Travis Scott hosted a concert that was attended live by 28 million players, with millions more watching live on social media. The track Scott premiered during the concert, which featured Kid Cudi, debuted at #1 on the *Billboard* Hot 100 charts a week later, was Cudi's first #1 track, and finished 2020 as the third-largest US debut of the year. In addition, several of the tracks Scott performed from his two-year-old *Astroworld* album returned to the *Billboard* charts after the concert. Eighteen months later, *Fortnite*'s official event video had accumulated nearly 200 million views on YouTube.

The multi-decade history of social virtual worlds, from MUDs to *Fortnite*, helps explain why the ideas of the Metaverse have recently shifted from science fiction and patents to the forefront of consumer and enterprise technology. We are now at the point when these experiences can appeal to hundreds of millions and their bounds are more about the human imagination than technical limitation.

In mid-2021, only weeks before Facebook unveiled its Metaverse intentions, Tim Sweeney, CEO and founder of *Fortnite* maker Epic Games, tweeted prerelease code from the company's 1998 game *Unreal*, adding that players "could go into portals and travel among user-run servers when Unreal 1 was released in 1998. I remember a moment where folks in the community had created a grotto map with no combat and were standing in a circle chatting. This style of play didn't last for long though."[9] A few minutes later, he added: "We've had metaverse aspirations for a very, very long time . . . but only in recent years have a critical mass of working pieces started coming together rapidly."[10]

This is the arc of all technological transformations. The mobile internet has existed since 1991, and was predicted long before. But it was only in the late 2000s that the requisite mix of wireless speeds, wireless devices, and wireless applications had advanced to the point where every adult in the developed world—and within a decade, most people on earth—would want and be able to afford a smartphone and broadband plan. This in turn led to a transformation of digital information services and human culture at large. Consider the following:

when instant messaging pioneer ICQ was acquired by internet giant AOL in 1998, it had 12 million users. A decade later, Facebook had over 100 million monthly users. By the end of 2021, Facebook had 3 billion monthly users, with some 2 billion using the service daily.

Some of this change, too, is a result of generational succession. For the first two or so years following the release of the iPad, it was common to see press reports and viral YouTube videos of infants and young children who would pick up an "analogue" magazine or book and try to "swipe" its nonexistent touchscreen. Today, those one-year-olds are eleven to twelve. A four-year-old in 2011 is now well on her way to adulthood. These media consumers are now spending their own money on content—and some are already creating content themselves. And while these previously unintelligible consumers now understand why adults found their futile efforts to pinch-to-zoom a piece of paper so comic, older generations are not much closer to understanding how the worldviews and preferences of the young differ from their own.

*Roblox* is the perfect case study of this phenomenon. The platform launched in 2006 and roughly a decade passed before it had much of an audience. Another three years went by before non-players really noticed the title (and those who did largely scoffed at its low-fidelity graphics). Two years later, it was one of the biggest media experiences in history. This 15-year timeline is partly a result of the technical improvements, but it's no coincidence that *Roblox*'s core users are the very children who grew up "iPad Native." The success of *Roblox*, in other words, required other technologies to influence how consumers thought, in addition to enabling it in the first place.

## The Coming Fight to Control the Metaverse (and You)

Over the past 70 years, "proto-Metaverses" have grown from text-based chats and MUDs to vivid networks of virtual worlds with populations and economies that rival small nations. This trajectory will continue in the decades to come, bringing more realism, diversity of experiences, participants, cultural influence, and value to virtual worlds. Eventually, a version of the Metaverse as imagined by Stephenson, Gibson, Baudrillard, and others will be realized.

There will be many wars for supremacy in and over this Metaverse. They will be fought between tech giants and insurgent start-ups through hardware, technical standards, and tools, as well as content, digital wallets, and virtual identities. This fight will be motivated by more than just revenue potential or the need to survive the "pivot to Metaverse."

In 2016, a year before his company's release of *Fortnite* and long before the term "Metaverse" entered public consciousness, Tim Sweeney told reporters: "This Metaverse is going to be far more pervasive and powerful than anything else. If one central company gains control of this, they will become more powerful than any government and be a God on Earth."*[11] It is easy to find such a statement hyperbolic. The provenance of the internet, however, suggests that it may not be.

The foundation of today's internet was built over several decades and through a variety of consortiums and informal working groups composed of government research labs, universities, and independent technologists and institutions. These mostly not-for-profit collectives typically focused on establishing open standards that would help them share information from one server to another, and in doing so make it easier to collaborate on future technologies, projects, and ideas.

The benefits of this approach were far-ranging. For example, anyone with an internet connection could build a website in minutes and at no cost using pure HTML, and even faster using a platform like GeoCities. A single version of this site was (or at least could be) accessed by every device, browser, and user connected to the internet. In addition, no user or developer needed to be disintermediated—they

---

* In its ruling for *Epic Games, Inc. v. Apple Inc.*, the district court wrote "[It] generally finds Mr. Sweeney's personal beliefs about the future of the metaverse are sincerely held" (*Epic Games, Inc. v. Apple Inc.*, U.S. District Court, Northern District of California, Case 4:20-cv-05640-YGR, Document 812, filed September 10, 2021).

could produce content for, and speak to, anyone they wanted. The use of common standards also meant that it was easier and cheaper to hire and work with outside vendors, integrate third-party software and apps, and repurpose code. The fact that so many of these standards were free and open-source meant that individual innovations often benefited the entire ecosystem, while placing competitive pressures on paid, proprietary standards, and helping to check the rent-seeking tendencies of platforms sitting between the web and its users (e.g., device manufacturers, operating systems, browsers, and ISPs).

Importantly, none of this prevented businesses from making a profit on the internet, deploying a paywall, or building proprietary technology. Rather, the "openness" of the internet enabled more companies to be built, in more areas, reaching more users, and achieving greater profits, while also preventing pre-internet giants (and, crucially, telecom companies) from controlling it. Openness is also why the internet is largely considered to have democratized information, and why the majority of the most valuable public companies in the world today were founded (or were reborn) in the internet era.

It's not difficult to imagine how different the internet would be if it had been created by multinational media conglomerates in order to sell widgets, serve ads, harvest user data for profits, or control users' end-to-end experience (something AT&T and AOL both tried but failed to pull off). Downloading a JPG could cost money, and a PNG could cost 50% more. Video calls might have only been possible through a broadband operator's own app or portal—and only to those who also had that same broadband provider (imagine something like, "Welcome to your Xfinity Browser™, click here for Xfinitybook™ or Xfinity-Calls™ powered by Zoom™; Sorry, 'Grandma' is not in our network, but for $2, you can still call her . . ."). Imagine if it took a year or a thousand dollars to make a website. Or if websites only worked in Internet Explorer or Chrome—and you had to pay a given browser an annual fee for the privilege of using it. Or maybe you would have to pay your broadband provider extra fees to read certain programming languages or use a given web technology (imagine, again, "This website requires Xfinity Premium with 3D"). When the United States sued Microsoft in 1998 for alleged antitrust violations, it centered its case on Microsoft's decision to bundle Internet Explorer, the company's proprietary web browser, with the Windows operating system (OS). Yet if a corporation had created the internet, is it conceivable that it would have even allowed a competing browser? If so, would it have allowed users to do whatever they wanted on these browsers, or access (and modify) whichever sites they chose?

A "corporate internet" is the current expectation for the Metaverse. The internet's nonprofit nature and early history stem from the fact that government research labs and universities were effectively the only institutions with the computational talent, resources, and ambitions to build a "network of networks," and few in the for-profit sector understood its commercial potential. None of this is true when it comes to the Metaverse. Instead, it is being pioneered and built by private businesses, for the explicit purpose of commerce, data collection, advertising, and the sale of virtual products.

What's more, the Metaverse is emerging at a time when the largest vertical and horizontal tech platforms have already established enormous influence over our lives, as well as the technologies and business models of the modern economy. This power partly reflects the profound feedback loops in the digital era. Metcalfe's Law, for example, states that the value of a communication network is proportional to the square of the number of its users, a relationship that helps to keep large social networks and services growing and presents a challenge to upstart competitors. Any business based on artificial intelligence or machine learning benefits from similar advantages as their datasets grow. The primary business models of the internet—advertising and software sales—are also scale-driven, as the companies that sell another ad slot or app encounter almost no incremental cost from doing so, and both advertisers and developers focus primarily on where consumers already are, rather than where they might be.

But to secure their user and developer bases while also expanding into new areas and blocking potential competitors, the tech giants have spent the past decade closing their ecosystems. They've done this by forcibly bundling together their many services, preventing users and developers from easily exporting their own data, shutting down various partner programs, and stymying (if not outright blocking) for-profit and even open standards which might threaten their hegemony. These maneuvers, mixed with the feedback loops that come from having comparatively more users, data, revenue, devices, etc., have effectively closed much of the internet. Today, a developer must essentially receive permission and provide payment. Users have little ownership of their online identity, data, or entitlements.

It is here that fears of a Metaverse dystopia seem fair, rather than alarmist. The very idea of the Metaverse means an ever-growing share of our lives, labor, leisure, time, wealth, happiness, and relationships will be spent inside virtual worlds, rather than just extended or aided through digital devices and software. It will be a parallel plane of existence for millions, if not billions, of people, that sits atop our digital and physical economies, and unites both. As a result, the companies that control these virtual worlds and their virtual atoms will likely be more dominant than those who lead in today's digital economy.

The Metaverse will also render more acute many of the hard problems of digital existence today, such as data rights, data security, misinformation and radicalization, platform power and regulation, abuse, and user happiness. The philosophies, culture, and priorities of the companies that lead in the Metaverse era, therefore, will help determine whether the future is better or worse than our current moment, rather than just more virtual or remunerative.

As the world's largest corporations and most ambitious start-ups pursue the Metaverse, it's essential that we—users, developers, consumers, and voters—understand that we have agency over our future and the ability to reset the status quo. Yes, the Metaverse can seem daunting and scary, but it also offers a chance to bring people closer together, to transform industries that have long resisted disruption and that must evolve, and to build a more equal global economy. This leads us to one of the most exciting aspects of the Metaverse: how poorly understood it is today.

## Chapter 2

# CONFUSION AND UNCERTAINTY

FOR ALL THE FASCINATION WITH THE METAVERSE, the term has no consensus definition or consistent description. Most industry leaders define it in the manner that fits their own worldviews and/or the capabilities of their companies.

For example, Microsoft's CEO Satya Nadella has described the Metaverse as a platform that turns the "entire world into an app canvas"[1] which could be augmented by cloud software and machine learning. No surprise, Microsoft already had a "technology stack"[2] which was a "natural fit" for the not-quite-here Metaverse and spanned the company's operating system Windows, cloud computing offering Azure, communications platform Microsoft Teams, augmented reality headset HoloLens, gaming platform Xbox, professional network LinkedIn, and Microsoft's own "Metaverses" including *Minecraft*, *Microsoft Flight Simulator,* and even the space-faring first-person shooter *Halo.*[3]

Mark Zuckerberg's articulation focused on immersive virtual reality* as well as social experiences that connect individuals who live far

---

* "Virtual reality applications" technically refers to computer-generated simulations of three-dimensional objects or environments with seemingly real, direct, or physical user interaction (J. D. N. Dionisio, W. G. Burns III, and R. Gilbert, "3D Virtual Worlds and the Metaverse: Current Status and Future Possibilities," *ACM Computing Surveys* 45, issue 3 [June 2013], http://dx.doi.org/10.1145/2480741 .2480751). In modern usage, it most commonly refers to immersive virtual reality,

apart. Notably, Facebook's Oculus division is the market leader in VR in both unit sales and investment, while its social network is the largest and most used globally. The *Washington Post* characterized Epic's vision of the Metaverse, meanwhile, as "an expansive, digitized communal space where users can mingle freely with brands and one another in ways that permit self-expression and spark joy . . . a kind of online playground where users could join friends to play a multiplayer game like Epic's 'Fortnite' one moment, watch a movie via Netflix the next and then bring their friends to test drive a new car that's crafted exactly the same in the real world as it would be in this virtual one. It would not be (in Sweeney's opinion) the manicured, ad-laden news feed presented by platforms like Facebook."[4]

In many cases, the Metaverse discourse showed that executives see the need to use the buzzword before they have any real handle on what it means overall, let alone for their business. In August 2021, Match Group, the owner of dating sites such as Tinder, Hinge, and OKCupid, said that its services would soon receive "augmented features, self-expression tools, conversational AI and a number of what we would consider metaverse elements, which have the element to transform the online meeting and getting-to-know-each-other process." No further details were provided, though presumably its Metaverse initiatives will involve virtual goods, currencies, avatars, and environments that facilitate romance.

After Chinese megacorporations Tencent, Alibaba, and ByteDance began positioning themselves as leaders in the vaguely defined but seemingly imminent Metaverse, their domestic competitors stumbled as they sought to explain how they, too, would become pioneers in this multi-trillion-dollar future. For instance, the head of investor relations at NetEase, another Chinese gaming giant, said on the company's Q3

---

in which a user's sense of sight and sound are fully placed within this environment, in contrast to viewing it on a device such as a TV, in which only parts of their senses are immersed in the environment.

2021 earnings call that "The metaverse is indeed the new buzzword everywhere today. But then, on the other hand, I think nobody has actually had firsthand experience in what it is. But at NetEase, we are technologically ready. We know how to accumulate the relevant know-how, the relevant skillsets when that day comes. So, I think when that day eventually comes, we'd probably be one of the fastest runners in the metaverse space."[5]

A week after Zuckerberg first detailed his Metaverse strategy, CNBC's Jim Cramer found himself the subject of online mockery after he struggled to explain the Metaverse to Wall Street investors.[6]

**Jim Cramer (JC):** You have to go to the Unity conference call first quarter, which really explains what the Metaverse is, which is the idea that you you're, you're, you're looking at basically you can be in Oculus, whatever. And you say, I like the way that person looks in that shirt. I want to order that shirt and it's or ultimately it's an NVIDIA uh, based on NVIDIA. And when I was out at NVIDIA with Jensen Huang, what happens? You could, it's conceivable. Okay. David, listen to me. Cause this is important.

**David Faber (DF):** I'm reading what the Zuckerberg had to say about it—

JC:—he didn't tell you nothing . . . no, he didn't!

DF:—"a persistent synchronous environment where we can be together, which I think is probably going to resemble some kind of a hybrid between the social platforms we've seen today, but an environment where you're embodied in it." That tells me what it is: it's The Holodeck.

JC:—It IS a hologram. It's like the idea—

DF:—it's like Star Trek—

JC:—ultimately, you could go into a room, let's say you're alone and you're a little lonely, okay? And you like classical music, but you go into the room and you say to the first person you see, "Do you think that you like to do you like the Mozart, you know, the Haffner?" And then the second person says, "Before you listen to Haffner, have you listened to Beethoven's ninth?" Let me tell you, these people don't exist. Okay?

DF:—Understood.

JC:—THAT'S the Metaverse.

While Cramer was obviously confused, much of the tech community continues to dispute key elements of the Metaverse. Some observers debate whether augmented reality is part of the Metaverse, or separate from it, and whether the Metaverse can only be experienced through immersive VR headsets or is just best experienced using such devices. To many in the crypto and blockchain community, the Metaverse is a decentralized version of today's internet—one in which users, not platforms, control its underlying systems, as well as their own data and virtual goods. Some important voices, such as former Oculus VR CTO John Carmack, argue that if the Metaverse is primarily operated by a single company, then it cannot be the Metaverse. Unity's CEO, John Riccitiello, doesn't subscribe to this belief, though he notes that the solution to the danger of a centrally controlled Metaverse are technologies such as Unity's cross-platform engine and services suite, which "pulls down the height of the wall of the walled garden." Facebook hasn't said whether or not the Metaverse can be privately operated, but the company does say that there can be only one Metaverse—just as there is "the internet," not "an internet" or "the internets." Microsoft and Roblox, conversely, talk about "Metaverses."

To the extent there is a common understanding of the Metaverse, it could be described as follows: a never-ending virtual world where everyone dresses up as comical avatars and competes in immersive VR games to win points, jumps into their favorite franchises, and acts out their most impossible fantasies. This was brought to life in Ernest Cline's *Ready Player One*, a 2011 novel considered to be a more mainstream, spiritual successor to Stephenson's *Snow Crash*, and which was adapted to film by Steven Spielberg in 2018. Like Stephenson, Cline never provided a clear definition of the Metaverse (or what he called "The Oasis"), but instead described it through what could be done and who one could be within it. This vision of the Metaverse is similar to how the average person understood the internet in the 1990s—it was "The Information Superhighway" or "World Wide Web," which we'd "surf" with our keyboards and "mouse"—just now in 3D. A quarter century later, it's obvious this conception of the internet was a poor and misleading way to describe what was to come.

The disagreement and confusion over the Metaverse, on top of its connection to partly dystopic sci-fi novels in which techno-capitalists rule two planes of human existence, result in a variety of critiques. Some argue the term represents little more than vapid marketing hype. Others wonder how the Metaverse will be any different from experiences such as *Second Life*, which have existed for decades and, though once expected to change the world, eventually faded from memory and was uninstalled from personal computers.

Some journalists have suggested that big tech's sudden interest in the nebulous idea of the Metaverse is actually an effort to avoid regulatory action.[7] Should governments around the world become convinced that a disruptive platform shift is imminent, this theory supposes then even the largest and most entrenched companies in history need not be broken up—free markets and insurgent competitors will do the work. Others have argued that, on the contrary, the Metaverse is being used by said insurgents so that regulators will open antitrust investigations into today's big tech leaders. One week before filing suit against Apple on antitrust grounds, Sweeney tweeted "Apple has outlawed the Metaverse," with the company's legal filings detailing how Apple's policies would prevent its emergence.[8] The federal judge presiding over the lawsuit seemed to buy into at least part of the "Metaverse as a regulatory strategy" theory, stating in court: "Let's be clear. Epic is here because, if relief is granted, it could turn the multibillion-dollar company into maybe a multitrillion-dollar company. They aren't doing it out of the kindness of their heart."[9] The judge also wrote that regarding Epic's lawsuit against Apple and

Google, "The record reveals two primary reasons motivating the action. First and foremost, Epic Games seeks a systematic change which would result in tremendous monetary gain and wealth. Second, [the lawsuit] is a mechanism to challenge the policies and practices of Apple and Google which are an impediment to Mr. Sweeney's vision of the oncoming metaverse."[10] Others have argued that CEOs are using the vaguely understood term to justify pet R&D projects that are years from public release, probably farther behind schedule, and of little interest to shareholders.

## Confusion as a Necessary Feature of Disruption

All new and particularly disruptive technologies deserve scrutiny and skepticism. But current debates about the Metaverse remain muddled because—at least thus far—the Metaverse is only a theory. It is an intangible idea, not a touchable product. As a result, it's difficult to falsify any specific claim, and inevitable that the Metaverse is understood within the context of a given company's own capabilities and preferences.

However, the sheer number of companies that see potential value in the Metaverse speaks to the size and diversity of the opportunity. What's more, debate over what the Metaverse is, how significant it might be, when it will arrive, how it will work, and the technological advances that will be required is exactly what produces the opportunity for widespread disruption. Far from disproving it, uncertainty and confusion are features of disruption.

Consider the internet. Wikipedia's description of the internet (which remains largely unchanged since the mid-2000s) goes as follows: "The global system of interconnected computer networks that uses the Internet Protocol Suite (TCP/IP) to communicate between networks and devices. It is a 'network of networks' that consists of private, public, academic, business, and government networks of local to global scope, linked by a broad array of electronic, wireless, and optical networking technologies. The internet carries a vast range of information resources and services, such as the inter-linked hypertext documents and applications of the World Wide Web (WWW), electronic mail, telephony, and file sharing."[11]

Wikipedia's summary addresses some of the internet's underlying technical standards and describes its scope as well as some of its use cases. The average person can read it today and easily map it to their personal usage and probably recognize why it's an effective definition. But even if you understood this definition in the 1990s—or even after Y2K—it didn't clearly explain what the future might look like. Even experts struggled to understand what to build on the Internet, let alone when to do so or through which technologies. The internet's potential and needs are obvious now, but at the time, almost no one had a cohesive, easily communicated, and correct vision of the future.

This confusion leads to a few common error types. Sometimes, emerging tech is seen as a trivial toy. In other cases, its potential is understood, but not its nature. Most often people misunderstand which specific technologies will thrive and why. On occasion, we get everything right except for the timing.

In 1998, Paul Krugman, who would win the Nobel Memorial Prize in Economic Sciences a decade later, wrote an (unintentionally) ironically titled article "Why Most Economists' Predictions Are Wrong" in which he stated: "The growth of the Internet will slow drastically, as the flaw in 'Metcalfe's law'—which states that the number of potential connections in a network is proportional to the square of the number of participants—becomes apparent: most people have nothing to say to each other! By 2005 or so, it will become clear that the Internet's impact on the economy has been no greater than the fax machine's."[12]

Krugman's prediction, which predated the dotcom crash, as well as the founding of companies such as Facebook, Tencent, and Pay-Pal, was quickly disproven. However, the internet's significance was debated for over a decade after his pronouncement. It wasn't until the mid-2010s, for example, that Hollywood accepted that the core of their businesses, not just low-cost, user-generated content such as YouTube videos and Snapchat Stories, would shift to the internet.

Even when the importance of the next platform is well understood, its technical premises, roles of related devices, and business models can remain unclear. In 1995, Microsoft founder and CEO Bill Gates wrote his famous "Internet Tidal Wave" memo, in which he explained that the internet was "crucial to every part of our business" and "the most important single development to come along since the IBM PC was introduced in 1981."[13] This rallying cry is considered the starting point of Microsoft's "Embrace, Extend, Extinguish" strategy, which the Department of Justice argued was part of the company's efforts to use its market power to catch up to and then eliminate market leaders in internet software and services.

Five years after Gates's memo, Microsoft launched its first mobile phone operating system. However, the company misread the dominant mobile form factor (the touchscreen); platform business model (app stores and services, rather than operating system sales); the role of the device (which became the primary computing device for most purchasers, rather than a secondary one); the extent of its appeal (everyone); its optimal price point ($500–$1,000); and its role (most functions, rather than just work and phone calls). As is well known today, Microsoft's mistakes came to a head beginning in 2007, when the first iPhone was released. When asked about the device's prospects, Microsoft's second-ever CEO, Steve Ballmer, infamously laughed and replied, "Five hundred dollars? Fully subsidized? With a plan? I said that is the most expensive phone in the world. . . . And it doesn't appeal to business customers because it doesn't have a keyboard. Which makes it not a very good email machine."[14] Microsoft's mobile operating system never recovered from the disruptive force of Apple's iPhone and iOS, nor of Google's Android, which targeted many of Microsoft's typical Windows manufacturers, such as Sony, Samsung, and Dell, but was free-to-license and even shared a portion of app store revenues with the device makers. By 2016, the majority of internet usage globally was via mobile computers. The next year—a decade after the first iPhone—Microsoft announced that it was discontinuing development of its Windows Phone.

Facebook, one of the biggest winners from the rise of the consumer internet, initially misjudged the mobile era too, but was able to corrects its mistakes before being displaced. Its mistake? Thinking that browsers, not apps, would be the dominant way to access the web.

Four years after Apple launched the iPhone's App Store, three years after Apple's famous "There's an app for that" ad campaign, and two years after *Sesame Street*, of all things, parodied that campaign, the social networking giant was still focused on browser-based experiences. While Facebook did technically release a mobile app the same day Apple released the App Store, and it quickly became the most popular way to access Facebook on a mobile device, this app was really just a "thin client" that loaded HTML inside a non-browser interface.

In mid-2012, Facebook finally relaunched its iOS app, which was "rebuilt from the ground up" to focus on device-specific code. Within a month, Mark Zuckerberg said that users were consuming "twice as many newsfeed stories" and that "the biggest mistake we made as a company was betting too much on HTML5. . . . We had to start over and rewrite everything to be native. We burned two years."[15] Ironically, Facebook's late shift to native apps is part of the reason the company is seen as a case study for successfully pivoting a business to mobile. Over the course of 2012, mobile's share of Facebook's total ad revenues surge from less than 5% to 23%—but this just demonstrates how much mobile revenue the company had lost out on by betting on HTML5 over the previous years. Facebook's delayed shift had other consequences in the form of missed opportunities and multi-billion-dollar bills. A decade after Facebook made its switch, the Facebook product with the most daily users is WhatsApp, which the company acquired in 2014 for nearly $20 billion. WhatsApp had been developed in 2009 specifically for app-based messaging on smartphones; at the time, Facebook had a nearly 350 million monthly user head start. Many on Wall Street also consider Instagram, the mobile-native social network that Facebook bought for $1 billion in the months prior to its relaunched iOS apps, to be its most valuable asset.

While Microsoft and Facebook made fundamental mistakes about the technologies of the future, many others failed because they bet on the right technology but before there was a market to support it. In the years before the dotcom crash, tens of billions of dollars were invested in building fiber optic networks across the United States. Due to the low marginal costs in laying extra capacity, many backers built considerably greater capacity than was needed—hoping that they would corner a regional market by providing enough capacity for all existing and future traffic. However, this was based on the faulty belief that internet traffic growth would increase exponentially in the years to come. Ultimately, it was common for less than 5% of fiber to be "lit," with the rest unused.

Today, the thousands of miles of "dark fiber" across America are a largely underappreciated enabler of the country's digital economy, silently helping content owners and consumers gain access to high-bandwidth, low-latency infrastructure at low prices. But in the years between the laying of these cables and the present day, many of those responsible went through bankruptcy. These include Metromedia Fiber Network, KPNQwest, 360networks, and, in one of the largest bankruptcies in US history, Global Crossing. Several other companies, such as Qwest and Williams Communications, barely escaped. Though ultimately felled by accounting fraud, the infamous collapses of WorldCom and Enron were exacerbated by multi-billion-dollar bets that demand for high-speed broadband would rapidly exceed supply. Enron was so convinced of the imminent and insatiable demand for highspeed data that in 1999 it unveiled plans to trade bandwidth futures like oil or silicon, assuming businesses would want to book capacity up to years in advance lest they encounter enormous swings in per bit delivery costs.

What makes technological transformation difficult to predict is the reality that it is caused not by any one invention, innovation, or individual, but instead requires many changes to come together. After a new technology is created, society and individual inventors respond to it, which leads to new behaviors and new products, which in turn lead to new use cases for the underlying technology, thereby inspiring additional behaviors and creations. And so on.

Recursive innovation is why even the biggest believers in the internet

20 years ago rarely predicted much about how it would be used today. The most accurate forecasts were typically platitudes such as "more of us will be online, more often, using more devices, and for more purposes," while the least accurate ones tended to be those that described exactly what we'd do online, when, where, how, and to what end. Certainly, few imagined a future in which entire generations would communicate primarily through emojis, tweets, or short filmed "Stories." Or where Reddit's stock investing forum, combined with free and easy investing via platforms such as Robinhood, would drive the rise of "You Only Live Once" trading strategies—which in turn saved companies such as GameStop and AMC Entertainment from COVID-19–driven bankruptcy. Or where 60-second-long TikTok remixes would define the *Billboard* charts, and with it, the soundtrack of our daily commutes. In 1950, IBM's product planning department reportedly spent the entire year "insisting that the market would never amount to more than about eighteen computers nationwide."[16] Why? Because the department could not imagine why anyone would need such devices, except to use the software and applications IBM was developing at the time.

Whether you're a Metaverse believer, skeptic, or somewhere in between, you should become comfortable with the fact that it is too early to know exactly what a "day in the life" might look and feel like when the Metaverse arrives. But the inability to precisely predict how we'll use it, and how it will change our daily life, is not a flaw. Rather, it is a prerequisite for the Metaverse's disruptive force. The only way to prepare for what is coming is to focus on the specific technologies and features that together comprise it. Put another way, we have to define the Metaverse.

*Chapter 3*

# A DEFINITION (FINALLY)

WITH THE IMPORTANT PRELIMINARIES BEHIND US, we can begin to talk concretely about what the Metaverse is. While there are competing definitions and a great deal of confusion, I believe it is possible to offer a clear, comprehensive, and useful definition of the term, even at this early point in the history of the Metaverse.

Here, then, is what I mean when I write and speak about the Metaverse: "A massively scaled and interoperable network of real-time rendered 3D virtual worlds that can be experienced synchronously and persistently by an effectively unlimited number of users with an individual sense of presence, and with continuity of data, such as identity, history, entitlements, objects, communications, and payments."

This chapter unpacks each element of this definition and in doing so explains not just the Metaverse, but how the Metaverse differs from today's internet, what will be needed to realize it, and when it might be achieved.

## Virtual Worlds

If there's any aspect of the Metaverse on which everyone—from believers to skeptics and even those barely familiar with the term—can agree, it's that it is based on virtual worlds. For decades, the primary reason to build a virtual world was for a video game, such as *The Legend of Zelda* or *Call of Duty*, or as part of a feature film, such as those of Disney's Pixar or for Warner Bros.' *The Matrix*. This is why the Metaverse is often misdescribed as a game or entertainment experience.

Virtual worlds refer to any computer-generated simulated environment. These environments can be in immersive 3D, 3D, 2.5D (also known as isometric 3D), 2D, layered atop the "real world" via augmented reality, or purely text-based, as in the game-like MUDs and non-game-like MUSHs of the 1970s. These worlds can have no individual user—as in the case of a Pixar film, or when virtually simulating an ecosphere for a biology class. In other cases, they might be limited to a single user, as when playing *Legend of Zelda*, or be shared with many others, as in *Call of Duty*. These users might affect and be affected by this virtual world through any number of devices, such as a keyboard, motion sensor, or even a camera that tracks their motion.

Stylistically, virtual worlds can reproduce the "real world" exactly (these are often called a "digital twin") or represent a fictionalized version of it (such as *Super Mario Odyssey*'s New Donk City, or the quarter-scale Manhattan of PlayStation's 2018 game *Marvel's Spider-Man*), or represent an altogether fictional reality in which the impossible is commonplace. The purpose of a virtual world can be "game-like," which is to say there is an objective such as winning, killing, scoring, defeating, or solving, or the purpose can be "non-game-like" with objectives such as educational or vocational training, commerce, socializing, meditation, fitness, and more.

Perhaps surprisingly, most of the growth and popularity in virtual worlds over the past decade has been in those which either lack or downplay game-like objectives. Consider the best-selling game made exclusively for the Nintendo Switch platform. You might guess that I'm referring to 2017's *The Legend of Zelda: Breath of the Wild* or *Super Mario Odyssey*, both of which are frequently thought of as among the greatest games ever made and part of the most popular video game franchises in history. But neither title wears the crown. Instead, the victor is *Animal Crossing: New Horizons,* which comes from a celebrated and popular franchise, has been available for purchase less than a third as long as the other two Nintendo titles, yet outsold them by nearly 40%. While *Animal Crossing: New Horizons* is nominally a game, its actual gameplay has often been likened to a virtual form of gardening. There are no explicit goals, least of all something to win. Instead, users gather and craft items on a tropical island, foster a community of anthropomorphic animals, and trade decorative wares and creations with other players.

In recent years, the biggest uptick in virtual world creation has been via worlds which have no "gameplay" whatsoever. For example, a digital twin of the Hong Kong International Airport was created using the popular game engine Unity—the purpose of the twin was to simulate the flow of passengers, the implications of maintenance issues or runway backups, and other events that would impact airport design choices and operational decision-making. In other cases, entire cities have been re-created and then connected to real-time data feeds for vehicular traffic, weather, and other civic services, such as police, fire, ambulance response. The goal of such a digital twin is to enable city planners to better understand the cities they manage and make more informed decisions about zoning, construction approvals, and more. For example, how would a new commercial mall affect travel times for emergency medical or police services? How might a specific building design adversely affect wind conditions, urban temperatures, or downtown light? Virtual worlds can prove an essential aid.

Virtual worlds can have a single or many different creators, they can be professional or amateur, for-profit or not-for-profit. However, their popularity has surged as the cost, difficulty, and time required to create them has plummeted, in turn leading to increased numbers of virtual worlds and greater diversity among and within them. *Adopt Me!*, a *Roblox*-based experience, was developed by only two independent and otherwise inexperienced people in the summer of 2017. Four years later, the game had nearly 2 million players at a single time (*The Legend of Zelda: Breath of the Wild* has sold roughly 25 million copies in its lifetime), and by the end of 2021, it had been played more than 30 billion times.

Some virtual worlds are fully persistent, which means everything that happens inside them is permanent. In other cases, the experience is reset for each player. More often, a virtual world operates somewhere in the middle. Consider the famous 2D sidescrolling game *Super Mario Bros.*, released in 1985 for the Nintendo Entertainment System. The first level lasts no longer than 400 seconds. If the player dies before then, they might have an extra life which enables them to retry it, but the level's virtual world will have been fully reset as though the player had never been there before—that is, all enemies that were killed are returned to life, and all items restored. However, *Super Mario Bros.* also allows some items to persist. A player who dies in level 3–4 retains the coins collected during prior levels, as well as their progress in the game—until they run out of all their lives, after which all data is reset.

Some virtual worlds are limited to a specific device or platform. Examples here include *Legend of Zelda: Breath of the Wild*, *Super Mario Odyssey*, and *Animal Crossing: New Horizons*, which are available exclusively on Nintendo's Switch. Others operate on several platforms, such as Nintendo's mobile games, which run on most Android and iOS devices, but not the Nintendo Switch or any other consoles. Some titles are considered fully cross-platform. In 2019 and 2020, *Fortnite* was available on all of the major gaming consoles (e.g., Nintendo's Switch, Microsoft's Xbox One, Sony's PlayStation 4), PC devices (i.e., those running Windows or Mac OS), as well as the top mobile platforms (iOS and Android).* This meant that a single player could access the title, their account, and their owned goods (for instance, a virtual backpack or outfit) from nearly any device. In other cases, titles are nominally available on multiple platforms, but the experiences are disconnected. *Call of Duty Mobile* and the PC/console-only *Call of Duty Warzone* share select account information and are both battle royale games with similar maps and mechanics,

---

* After Epic Games sued Apple in August 2020, Apple removed *Fortnite* from its App Store, thereby making it impossible for users to play the game on iOS devices.

but are otherwise different games and players in one virtual world cannot play against players in the other.

As with the real world, the governance models of virtual worlds vary greatly. Most are centrally controlled by the person or group that developed and operates the world, which means they have unilateral control over its economy, policies, and users. In other instances, users self-govern through various forms of democracy. Some blockchain-based games aspire to operate as close to autonomously as possible after launch.

# 3D

Although virtual worlds come in many dimensions, "3D" is a critical specification for the Metaverse. Without 3D, we might as well be describing the current internet. Message boards, chat services, website builders, image platforms, and interconnected networks of content have been around and popular for decades, after all.

3D is necessary not just because it signals something new. Metaverse theorists argue that 3D environments are required in order to make possible the transition of human culture and labor from the physical world to the digital one. For example, Mark Zuckerberg has claimed that 3D is an inherently more intuitive interaction model for humans than 2D websites, apps, and video calls—especially in social use cases. Certainly, humans did not evolve for thousands of years to use a flat touchscreen.

We must also consider the nature of online communities and experiences over the last several decades. In the 1980s and early 1990s, the internet was mostly text-based. An online user represented their identity via username or email address, and a written profile, and expressed themselves via chat rooms and message boards. In the late 1990s and early 2000s, PCs became capable of storing larger files, while internet speeds made it practical to upload and download them. Accordingly, most internet users began to represent themselves online through display/profile pictures, as well as personal websites that included a handful of low-resolution images and sometimes even audio clips. Eventually, this led to the emergence of the first mainstream social networks, such as MySpace and Facebook. By the late 2000s and early 2010s, altogether new forms of online socializing began to emerge. Gone were the days of infrequently updated personal blogs or Facebook pages comprising a single cover photo and a string of old, text-only status updates. Instead, users expressed themselves through a near-constant stream of high-resolution photos and even videos—many of which were taken on the go and for no purpose other than to share what they were doing, eating, or thinking at a given moment. Again, this was led by altogether new social media networks such as YouTube, Instagram, Snapchat, and TikTok.

This history provides a few lessons. First, humans seek out digital models that most closely represent the world as they experience it—richly detailed, mixing audio and video, and with a sense of being "live" rather than static or outdated. Second, as our online experiences become more "real," we place more of our real lives online, live more of our lives online, and human culture overall becomes more affected by the online world. Third, the leading indicator for this change is typically new social apps, which, more often than not, are first embraced by younger generations. Collectively, these lessons seem to support the notion that the next great step for the internet is 3D.

If this is indeed the case, we can imagine how a "3D internet" might finally disrupt industries that have largely resisted digital disruption. For decades, futurists have predicted that education, most notably post-secondary education and vocational training, would be partly displaced by remote, online schooling. Instead, the cost of traditional, in person education has continued to rise (and at orders of magnitude above the average rate of inflation), while applications to colleges and universities continue to surge—even though the experience remains mostly unchanged. None of the most prestigious schools in the world have even tried to launch remote education programs that aspire to match the quality or imprimatur of their in-person equivalent, in part because employers seem unlikely to recognize them as such. And for millions of parents worldwide, the COVID-19 pandemic was a lesson on the inadequacy of children learning alone via 2D touchscreen. Many imagine that the improvements to 3D virtual worlds and simulations, as well as VR and AR headsets, will fundamentally reshape our pedagogical practices. Students from around the world will be able to strap into a virtual classroom, sit alongside their peers while making eye contact with their teacher, then shrink down to blood cells which travel through a human circulatory system, after which these previously 15-micrometer-tall students re-enlarge and dissect a virtual cat.

It is important to emphasize that while the Metaverse should be understood as a 3D experience, this does not mean that everything inside the Metaverse will be in 3D. Many people will play 2D games inside the Metaverse, or use the Metaverse to access software and applications that they then experience using mobile-era devices and interfaces. In addition, the advent of the 3D Metaverse does not mean that the entirety of the internet and computing at large will transition to 3D; the mobile internet era started more than a decade and a half ago, and yet many still use non-mobile devices and networks. Moreover, data transmitted between two mobile devices is still primarily transmitted through wired (i.e., underground) internet infrastructure. And despite the proliferation of the internet over the past 40 years, there are still offline networks and networks using proprietary protocols. However, it is 3D that enables so many new experiences to be built on the internet—and that creates the extraordinary technical challenges described next.

I should also note that no part of the Metaverse requires an immersive virtual reality or VR headset. These may eventually be the most popular way to experience the Metaverse, but immersive virtual reality is just one way to access it. Arguing that immersive VR is a requirement for the Metaverse is similar to arguing that the mobile internet can only be accessed via apps, thereby excluding mobile browsers. In truth, we don't even need a screen to access mobile data networks and mobile content, as is often the case with vehicular tracking devices, select headphones, and countless machine-to-machine and internet of things (IoT) devices and sensors. (The Metaverse won't require screens either, by the way—more on this in Chapter 9.)

# Real-Time Rendered

Rendering is the process of generating a 2D or 3D object or environment using a computer program. The goal of this program is to "solve" an equation made up of many different inputs, data, and rules that determine what should be rendered (that is, visualized) and when, and by using various computing resources, such as a graphics processing unit (or GPU) and central processing unit (CPU). As is the case with any math problem, an increase in the resources available to solve it (in this case, time, the number of CPUs/GPUs, and processing power) means that more complex equations can be tackled, and more detail provided in the solution.

Take the 2013 film *Monsters University*. Even when using an industrial grade computing processor, it would have taken an average of 29 hours for each of the film's 120,000-plus frames to be rendered. In total, that would have meant more than two years just to render the entire movie once, assuming not a single render was ever replaced or scene changed. With this challenge in mind, Pixar built a data center of 2,000 conjoined industrial-grade computers with a combined 24,000 cores that, when fully assigned, could render a frame in roughly seven seconds.[1] Most companies, of course, can't afford such a powerful supercomputer and therefore spend more time waiting. Many architectural and design firms, for example, need to wait overnight to render a highly detailed model.

Prioritizing visual fidelity is sensible if you're creating a Hollywood blockbuster that will be shown on an IMAX screen, or when you're selling a multi-million-dollar building renovation. However, experiences set in virtual worlds require *real-time* rendering. Without real-time rendering, the size and visuals of virtual worlds would be severely constrained, as would the number of participating users and the options available to each user. Why? Because experiencing an immersive environment through pre-rendered images requires every possible sequence to have been pre-made—just as a choose-your-own-adventure novel can offer only a handful of choices, rather than infinite ones. In other words, the cost of greater visuals is less functionality and agency.

Compare, for example, navigating the Roman Colosseum in a video game versus doing the same on Google Street View. Both provide 360-degree views and multiple dimensions of movement (look up or down, move left or right, backward or forward), but the former severely limits one's choices—and if you decide to look closely at a given stone, all you can do is zoom into an image not designed for such scrutiny. It will be blurry, and the view angle is fixed.

Although real-time rendering enables a virtual world to be "alive" and respond to input from a user (or a group of users, for that matter), it means that a minimum of 30, and ideally 120 frames, must be rendered each second. This constraint necessarily affects which and how much hardware is used and for how many cycles, and thereby the complexity of what's rendered. As you might expect, immersive 3D requires far more intensive computing power than 2D. And just as the average architectural firm cannot contend with the supercomputers built by a Disney subsidiary, the average user can't afford the GPUs or CPUs used by a corporation.

## Interoperable Network

Central to most visions of the Metaverse is the user's ability to take her virtual "content," such as an avatar or a backpack, from one virtual world to another, where it might also be changed, sold, or remixed with other goods. For example, if I buy an outfit in *Minecraft* I might then wear it in *Roblox*, or perhaps a hat I purchased in *Minecraft* would be paired with a sweater I won in *Roblox* while attending a virtual sporting match developed and operated by FIFA. And if attendees of the match received an exclusive item at this event, they could take it with them from that environment into others, and even sell it on third party platforms as though it were an original 1969 Woodstock T-shirt.

In addition, the Metaverse should make it so that wherever a user goes or whatever they choose to do, their achievements, history, and even finances are recognized across multitudes of virtual worlds, as well as the real one. The closest analogues are the international passport system, local market credit scores, and the national identification systems (such as social security numbers).

To realize this vision, virtual worlds must first be "interoperable," a term that refers to the ability for computer systems or software to exchange and make use of information sent from one another.

The most significant example of interoperability is the internet, which enables countless independent, heterogeneous, and autonomous networks can safely, reliably, and comprehensibly exchange information globally. All of this is made possible by the adoption of the Internet Protocol Suite (TCP/IP), a set of communications protocols that tell disparate networks how data should be packetized, addressed, transmitted, routed, and received. This suite is managed by the Internet Engineering Task Force (IETF), a nonprofit open standards group established in 1986 under the US federal government (it has since become a fully independent and global body).

The establishment of TCP/IP did not alone produce the globally interoperable internet as we know it today. We say "the internet," rather than "an internet," and choose to use "the internet" over no practical alternatives, because nearly every computer network globally, from small-to-medium businesses and broadband providers, as well as device manufacturers and software companies, voluntarily embraced the Internet Protocol Suite.

In addition, new working bodies were established to ensure that, no matter how large and decentralized the internet and World Wide Web might become, it would continue to interoperate. These bodies managed the assignment and expansion of top-level hierarchical web domains (.com, .org, .edu), as well as IP addresses, which distinctly identify individual devices on the internet, the Uniform Resource

Locator (or URL), which specifies the location of a given resource on a computer network, and HTML.

Also important was the establishment of common standards for files on the internet (e.g., JPEG for digital images and MP3 for digital audio), common systems for presenting information on the internet that were built on linkages between different websites, webpages, and web content (such as HTML), and browser engines that could render this information (Apple's WebKit). In most cases, several competing standards were established, but technical solutions emerged to convert from one another (for example, a JPEG to a PNG). Due to the openness of the early web, most of these alternatives were open sourced and sought the broadest possible compatibility. Today, a photo taken on an iPhone can easily be uploaded to Facebook, then downloaded from Facebook to Google Drive, then posted to an Amazon review.

The internet demonstrates the scope of the systems, technical standards, and conventions required to establish, maintain, and scale interoperability across heterogenous applications, networks, devices, operating systems, languages, domains, countries, and more. Yet far more still will be needed to realize visions of an interoperable network of virtual worlds.

Almost all the most popular virtual worlds today use their own different rendering engines (many publishers operate several across their titles), save their objects, textures, and player data into entirely different file formats and with only the information that they expect to need, and have no systems through which to even try to share data to other virtual worlds. As a result, existing virtual worlds have no clear way to find and recognize one another, nor do they have a common language in which they can coimmunicate, let alone coherently, securely, and comprehensively.

This isolation and fragmentation stems from the fact that today's virtual worlds, and their builders, never designed their systems or experiences to be interoperable. Instead, they were intended to be closed experiences with controlled economies—and optimized accordingly.

There is no obvious or fast way to establish standards and solutions. Consider, for example, the idea of an "interoperable avatar." It's relatively easy for developers to agree on the definition of an image and how to present it, and as a static 2D unit of content made up of individually colored pixels, the process of converting one image filetype (say, PNG) to another (JPEG) is straightforward. However, 3D avatars are a more complex question. Is an avatar a complete 3D person with an outfit, or is it made up of a body avatar plus an outfit? If the latter, how many articles of clothing are they wearing and what defines a shirt versus a jacket that goes over shirts? Which parts of an avatar can be recolored? Which parts must be recolored together (is a sleeve separated from a shirt)? Is an avatar's head a complete object, or is it a description of dozens of sub-elements like individual eyes (with their own retinas), eyelashes, noses, freckles, and so on. In addition, users expect an anthropomorphic jellyfish avatar and a box-like android to move in different ways. The same applies to objects. If a tattoo is a placed on the avatar's neck, it should stay fixed to their skin irrespective of any movement they make. A tie hung around that neck, however, should move with (and also interact with) the avatar as it moves. And it should move differently than a seashell necklace, which should also move differently than a feather necklace. Just sharing the dimensions and visual detail of an avatar is not sufficient. Developers need to understand, and agree on, how they work.

Even if new standards are agreed upon and improved, developers will need code that can properly interpret, modify, and approve third-party virtual goods. If *Call of Duty* wants to import an avatar from *Fortnite*, it will likely want to restyle the avatar to fit *Call of Duty's* gritty realism. To this end, it might want to reject those that cannot make sense in its virtual world, such as *Fortnite's* famous Peely skin, a giant anthropomorphic banana (which probably can't fit inside *Call of Duty's* cars or doorframes).

Other problems need to be resolved, too. If a user purchases a virtual good in one virtual world, but then uses it in many others, where is their ownership record managed and how is this record updated? How does another virtual world request this good on behalf of its supposed owner and then validate that the user owns it? How is monetization managed? Not only are unchangeable images and audio files simpler than 3D goods, but we can send copies of them between computers and networks and, critically, do not need to control how they're used thereafter and who has the right to use them.

And the above just concerns virtual objects. There are additional and largely unique challenges involved in interoperable identifies, digital communications, and especially payments.

What's more, we need the standards that are selected to be highly efficient. Consider, as an example, the GIF format. Though it's popular, it's awful technically. GIF images are typically very heavy (that is, their file size is relatively large) despite having compressed the source video file to the point that many individual frames are discarded and the remaining frames having lost much of their visual detail. The MP4 format, conversely, is typically five to ten times lighter and provides far greater video clarity and detail. The comparatively widespread use of GIF has therefore led to extra bandwidth use, more time waiting for files to load, and worse experiences overall. This may not seem like a terrible outcome, but as I'll discuss later in this book, the computational, network, and hardware demands of the Metaverse will be unprecedented. And 3D virtual objects are far heavier, and likely more important, than an image file. Which formats are selected will thus have a profound impact on what is possible, on which devices, and when.

The process of standardization is complicated, messy, and long. It is really a business and human problem masquerading as a technological one. Standards, unlike the laws of physics, are established through consensus, not discovery. Forming consensus often requires concessions that leave no party happy, which can then result in "forks" as different factions break off. Still, the process is never over. New standards are constantly emerging, with old ones updated and sometimes deprecated (we are slowly moving away from GIF). That the 3D standardization process is beginning decades after virtual worlds first emerged, and with trillions of dollars at stake, will make this even harder.

Pointing to these challenges, some argue that it is unlikely that "the Metaverse" will ever happen. Instead, there will be many competing networks of virtual worlds. But this is not an unfamiliar position. From the 1970s through to the early 1990s, there was also constant debate as to whether a common internetworking standard would be established (this period is known as the "Protocol Wars"). Most expected the world and its networks would be fragmented across a handful of proprietary networking stacks that spoke only to select outside networks and only for specific purposes.

In hindsight, the value of a single integrated internet is obvious. Without it, 20% of the world economy would not be "digital" today (nor much of the remainder digitally powered). And while not every company has benefited from openness and interoperability, most businesses and users have. Accordingly, the driving force behind interoperability is unlikely to be a given visionary voice or newly introduced technology, but instead will be economics. And the means of leveraging economics to the greatest degree will be common standards that will enhance the Metaverse economy by attracting more users and more developers, which will lead to better experiences, which in turn will be cheaper to make and more profitable to operate, thereby driving greater investment. It isn't necessary for all parties to embrace common standards, so long as economic gravity is allowed to do its work. Those who do will grow and those who don't will face constraints.

It's for this reason why it's so important to understand how the interoperability standards of the Metaverse will be established. The leaders here will have extraordinary soft power as this next-generation internet exists. In many ways, they will decide the rules of physics, and when, how, and why they will be updated.

## Massively Scaled

For "the internet" to be "the internet", we generally accept that it has to have a seemingly infinite number of websites and pages. It can't, for example, just be a handful of portals owned by a few developers. The Metaverse is similar. It must have a massively scaled number of virtual worlds if it is to be "the Metaverse." Otherwise, it is more like a digital theme park—a destination with a handful of carefully curated attractions and experiences that can never be as diverse as, or contend with, the outside (real) world.

Unpacking the etymology of the term "Metaverse" is helpful here. Stephenson's neologism comes from the Greek prefix "meta" and the stem "verse," a backformation of the word "universe." In English, "meta" roughly translates to "beyond" or "which transcends" the word that follows. For example, metadata is data that describes data, while metaphysics refers to a branch of philosophy "of being, identity and change, space and time, causality, necessity and possibility," rather than the study of "matter, its fundamental constituents, its motion and behavior through space and time, and the related entities of energy and force."[2] In combination, the "meta" and "verse" is intended to be a unifying layer that sits above and across all individual, computer generated "universes," as well as the real world, just as the universe contains, by some estimates, 70 quintillion planets.

Furthermore, within the Metaverse, there might be "metagalaxies," a collection of virtual worlds that all operate under a single authority and that are clearly connected by a visual layer. Under this definition, *Roblox* would be a Metagalaxy, while *Adopt Me!* would be a virtual world. Why? Because *Roblox* is a network of millions of different virtual worlds, one of which is *Adopt Me!*, but *Roblox* does not contain all virtual worlds (which would make it the Metaverse). Notably, individual virtual worlds might themselves have specific sub-regions, just as networks on the internet have their own sub-networks, and the earth has continents, often comprising many nations, which can be further divided into states and provinces, each containing cities, counties, and so on.

One way to think about a Metagalaxy is to think of Facebook's role in the internet. Facebook is obviously not the internet, but it is a collection of tightly integrated Facebook pages and profiles. In a simplified sense, it's today's version of a 2D Metagalaxy. Analogy also allows us to consider the likely extent of Metaverse interoperability. In today's universe, not all goods can travel everywhere. We could bring a guitar to Venus, but it would be immediately crushed; we could technically bring an Ohio farm to the moon, but it would be impractical. On earth, most human-made objects can be brought to most human-made places, however, we have various social, economic, cultural, and safety limitations which can get in the way of such efforts.

Growth in the number of virtual worlds should drive increased usage of virtual worlds. Some leaders within the virtual worlds space, such as Tim Sweeney, believe that eventually, every company will need to operate their own virtual worlds, both as standalone planets and as part of leading virtual world platforms such as *Fortnite* and *Minecraft*. As Sweeney has put it, "just as every company a few decades ago created a webpage, and then at some point every company created a Facebook page."

## Persistence

Earlier, I discussed the idea of persistence in a virtual world. Almost no current games demonstrate full persistence. Instead, they run for a finite period before resetting part or all of their virtual worlds. Consider the hit games *Fortnite* and *Free Fire*. Throughout a match, players build or destroy various structures, set fire to forests, or kill wildlife, but after roughly 20 to 25 minutes, the map effectively "ends" and is discarded by Epic Games and Garena—never to be re-experienced by a player, even if they retain items won or unlocked during that match. In fact, even within a given match, the virtual world discards data, such as a bullet mark on an indestructible rock, which might "unload" after 30 seconds in order to reduce render complexity.

Not all virtual worlds reset like a *Fortnite* match. *World of Warcraft*, for example, runs continuously. However, it's still wrong to say its virtual world fully persists. If a player enters a specific part of the *World of Warcraft* map, defeats their enemies, leaves, and returns, they'll more often than not find that those enemies have respawned. An in-game tradesman who sold a player a rare item only a day earlier might offer them a second as though it would be their first. Only when a large update is made by the developer, in this case Activision Blizzard, might a virtual world change. The players cannot themselves affect whether the consequences of a given choice or event endures indefinitely. The only thing that persists is the player's memory, and their record of having defeated an enemy or bought an item.

The challenge of persistence in virtual worlds can be a bit difficult to grasp because we don't encounter this problem in the real world. If you cut down a physical tree, it is gone irrespective of whether you personally remember cutting it down, and no matter how many other trees and activities Mother Earth is tracking. With a virtual tree, your device and the server which manages it must actively decide whether to retain this information, render it, and share it with others. And if these computers choose to do so, there are additional questions of detail—is the tree just "gone," or is it now felled on the ground? Should players see which side it was chopped from, or just that it was generically cut? And does it "biodegrade"? If so, how—generically, or in response to its local environments? The more information that persists, the greater the computational needs and the less memory and power that is available for other activities.

The best example of the computation-persistence interplay comes via the game *EVE Online*. Though not nearly as famous as other "proto-Metaverses" from the early 2000s, such as *Second Life*, nor newer ones such as *Roblox*, *EVE Online* is a marvel. With the exception of the occasional downtime for troubleshooting and updates, *EVE Online* has operated continuously and persistently since launching in 2003. And unlike games like *Fortnite*, which fragments its tens of millions of players into 20- to 30-minute matches of 12 to 150 players, *EVE Online* places its hundreds of thousands of monthly users into a single, shared virtual world that spans nearly 8,000 star systems and nearly 70,000 planets.

Behind *EVE Online*'s extraordinary virtual world is an innovative systems architecture—but also (and mostly) brilliant creative design.

The virtual world of *EVE Online* is essentially just empty three-dimensional space with wallpaper backgrounds that look like a galaxy. Users cannot truly visit a planet, with activities such as mining more akin to setting up a wireless router than constructing a virtual rig. As such, the game's persistence is mostly about managing a relatively modest set of entitlements (a player's ships and resources, for example) and related locational data. This means less computational work for CCP Games' servers, and for its users, whose devices need not render a changed world, just a few objects in it. Recall that complexity is the enemy of real-time rendering.

Furthermore, very little happens in *EVE Online* on a daily, quarterly, or even yearly basis. This is because the goal of *EVE Online*, to the extent one exists, is for various factions of players to conquer planets, systems, and galaxies. This is achieved primarily through the establishment of corporations, formation of alliances, and the strategic positioning of fleets. To this end, much of *EVE Online* actually takes place in the "real world," via third-party messaging application and emails, and not even on CCP's servers. Users have spent years planning attacks, going undercover with the enemy guilds in order to later betray them, and creating enormous personal networks that trade resources and construct new ships. While large-scale battles do happen, they're remarkably rare—and involve the destruction of assets in the virtual world (for example, ships) rather than the virtual world itself. The former is far easier for a processor to manage than the latter—just as throwing a garden plant in the garbage is easier than understanding how it'll affect the garden's ecosystem.

What makes *EVE Online* such an extraordinary example is how profoundly complex it is—both technically and sociologically—yet at the same time how limited compared to most visions of the Metaverse. In Stephenson's *Snow Crash*, the Metaverse is an enormous, planet-sized, and richly detailed virtual world with a nearly infinite number of unique businesses, places to visit, activities to do, things to buy, and people to meet. Nearly everything and anything done by any user, at any time, can persist forever. This applies not just to the virtual world, but to the individual items in it. Our avatars and virtual sneakers would wear with use and forever reflect their damage. And per the principles of interoperability, these modifications would persist wherever we go.

The amount of data that must be read, written, synchronized (more on this below), and rendered to create and sustain this experience is not just unprecedented—it is far beyond anything possible today. However, the literal version of Stephenson's Metaverse may not even be desirable. He imagined individuals waking up in the Metaverse inside their virtual homes, then walking or taking a train to a virtual bar. While skeuomorphism* often has utility, "The Street" as a single unifying layer for everything in the virtual world likely does not. Most participants in the Metaverse would rather teleport from destination to destination.

Fortunately, it is far easier to manage the persistence of a user's data (i.e., what they own and have done) across various worlds and over time, rather than the persistence of every user's most minute contributions to a planet-sized world. The model also more closely reflects the internet as it exists today—and probably our preferred interaction models, too. On the web, we often navigate directly to a webpage, such as a specific document in Google Docs or a video on YouTube. We don't start at some sort of "internet homepage," then click through to Google.com, then navigate to the appropriate product page, and so on.

Furthermore, the internet persists irrespective of any one site, platform, or top-level domain such as ".com." Should one site, or even many sites, cease to exist, content might be lost but the internet, as a whole, would persist. Much of a user's data, such as cookies or an IP address, not to mention the content they've created, can exist without a given website, browser, devices, platform, or service. If a virtual world goes offline, resets, or shuts down, however, it is almost as though it never existed for the player. Even if it continues to operate, the moment a player stops playing within a world, the virtual goods they own, their history and achievements, and even parts of their social graph are likely lost. This is less of a problem when virtual worlds are "games," but for human society to shift in a meaningful way into virtual spaces (i.e., for education, work, healthcare), what we do in these spaces must reliably endure, just as our grade school reports and baseball trophies

---

* "Skeuomorphism" refers to a technique used in graphical design in which interfaces are designed to mimic their real-world counterparts. For example, the iPhone's first "Notes" app involved typing on yellow paper with red lines, just like the common notepad.

do. To philosophers including John Locke, identity is better under-stood as continuity of memory. If so, then we can never have a virtual one as long as everything we do and have done is forgotten.

Increasing persistence within individual virtual worlds will neverthe-less be essential to the growth of the Metaverse. As I'll discuss through-out the rest of this book, many of the design ideas that have become popular over the past five years are not new, but rather newly possible. As such, we may currently struggle to figure out why *World of Warcraft* might need to forever remember a user's exact footprints in fresh snow, but odds are some designer will eventually figure out the answer and not long after, it will become a core feature of many games. Until then, the virtual worlds most in need of persistence are likely those based around virtual real estate, or tied to physical spaces. For example, we expect that "digital twins" should be frequently updated to reflect changes to their real-world counterpart, and that virtual-only real estate platforms would not "forget" about new art or décor added to a given room.

## Synchronous

We don't want virtual worlds in the Metaverse to merely persist or respond to us in real time. We also want them to be *shared* experiences.

For this to work, every participant in a virtual world must have an internet connection capable of transmitting large volumes of data in a given time ("high bandwidth"), as well as a low latency ("fast") and continuous* (sustained and uninterrupted) connection to a virtual world's server (both to and from).

This might not seem like an outlandish requirement. After all, tens of millions of homes are probably streaming high-definition video at this moment, while much of the global economy ran via live and synchronous video conference software throughout the COVID-19

---

* This is often referred to as a "persistent" connection, but in the interest of dif-ferentiating it from the persistence of a virtual world, I'll use the term "continu-ous" here.

pandemic. And broadband providers continue to boast about—and deliver—improvements in bandwidth and latency, with internet outages becoming less common each day.

However, synchronous online experiences are perhaps the greatest constraint facing the Metaverse today—and the one that is hardest to solve. Simply put, the internet was not designed for synchronous shared experiences. It was designed, instead, to allow for the sharing of static copies of messages and files from one party to another (namely research labs and universities that accessed them one at a time). Though this sounds impossibly limiting, it works pretty well for almost all online experiences today—specifically because almost none require continuous connectivity to feel live, or, well, continuous!

When a user believes they're surfing a live webpage, such as their constantly updating Facebook Newsfeed, or the Live Election feed from the *New York Times*, they're really just receiving frequently updated pages. What's actually happening is the following. To start, that user's device is making a request to Facebook or the *Times*' server, either via a browser or app. The server then processes the request, and sends back the appropriate content. This content includes code that requests updates from the server on a given interval (say, every 5 or 60 seconds). Furthermore, every one of these transmissions (from the user's device or that of the relevant server) might travel across a different set of networks to reach its recipient. While this feels like a live, continuous, and two-way connection, it's actually just batches of one-way, varyingly routed, and non-live data packets. The same model applies to what we call "instant messaging" applications. Users, and the servers between them, are really just pushing fixed data to one another, while frequently pinging for information requests (sending a message or sending a read receipt).

Even Netflix operates on a noncontinuous basis, even though the term "streaming" and target experience—uninterrupted playback—suggest otherwise. In truth, the company's servers are sending users distinct batches of data, many of which travel through different network paths from the server to that user. Netflix is often even pushing content to the user before it's needed—such as an extra 30 seconds.

Should a temporary delivery error occur (say, a specific pathway is congested, or the user briefly loses their Wi-Fi connection), the video will continue to play. The result of Netflix's approach is delivery which feels continuous, but only because it isn't delivered as such.

Netflix has other tricks, too. For example, the company receives video files anywhere from months to hours before they're made available to audiences. This gives the company a window during which it can perform extensive, machine learning–powered analysis that enables them to shrink (or "compress") file sizes by analyzing frame data to determine what information can be discarded. Specifically, the company's algorithms will "watch" a scene with blue skies and decide that, if a viewer's internet bandwidth suddenly drops, 500 different shades of blue can be simplified to 200, or 50, or 25. The streamer's analytics even do this on a contextual basis—recognizing that scenes of dialogue can tolerate more compression than those of faster-paced action. In addition, Netflix will pre-load content at local nodes. When you ask for the newest episode of *Stranger Things*, it's actually only a few blocks away and therefore arrives right away.

The approaches used above only work because Netflix is a non-synchronous experience; you can't "pre-do" anything for content that is being produced live. This is why live video streams, such as those of CNN or Twitch, are substantially less reliable than on demand streams from Netflix or HBO Max. But even live streamers have their own tricks. For example, transmission is typically delayed by two to thirty seconds, which means there's still the opportunity to pre-send content in case of temporary congestion. Commercial breaks can also be used by both the content provider's server, or the user, to reset the connection in case the prior one proved unreliable. Most live video requires only a one-way continuous connection—for instance, from CNN's server to the user. Sometimes there is a two-way connection, as in the case of a Twitch chat, but only a sparse amount of data is being shared (the chat itself) and it's not of critical importance—as it does not directly affect what's happening in the video (remember, it probably happened two to thirty seconds earlier).

Overall, very few online experiences require high bandwidth, low latency, and continuous connectivity other than real-time rendered, multiuser virtual worlds. Most experiences just need one or, at most, two of these elements. High-frequency stock traders (and especially high-frequency trading algorithms) want the lowest possible delivery times, as this can be the difference between buying or selling a security at a profit or a loss. However, the orders themselves are basic and light-weight, and don't require a continuous server connection.

The major exception is videoconferencing software such as Zoom, Google Meet, or Microsoft Teams, which involves many people receiving and sending high-resolution video files, all at once, and participating in a shared experience. However, these experiences are only possible through software solutions that don't really work for real-time rendered virtual worlds with many participants.

Think back to your last Zoom call. Every now and then, a few packets likely arrived too late or perhaps not at all, meaning you never heard a word or two—or perhaps, a few of your words were never heard by others on the call. Despite this, odds are you or your listeners still understood what was being said and the call could proceed. Maybe you temporarily lost, but then quickly regained, connectivity. Zoom can send you the packets you missed, then speed up playback and edit out pauses in order to "catch you up" to being "live." It's possible you lost your connection altogether, either due to a problem with your local network or through a problem encountered somewhere between your local network and a remote Zoom server. If this happened, you probably rejoined without anyone knowing you left—and if they did, it's unlikely your absence was disruptive. This is because videoconferences are shared experiences that focus on a single person, rather than a shared one that is led by many users working together. What if you were the speaker? The good news here is that the call could continue well enough without you, with either another participant piping up or everyone waiting for you to rejoin. If at any time network congestion meant that you or others simply could not hear or see what was happening, Zoom will stop uploading or downloading video from various members of the call, in order to prioritize what mattered most: audio. Or alternatively, the call might have been disrupted by varying latency—that is, different members of the call were receiving "live" video and audio a quarter, half, or even full second behind or in front of one another—resulting in struggles to take turns speaking and constant interruptions. Eventually, your call probably figured out how to manage this. Everyone just needs a little patience.

Virtual worlds have higher performance requirements and are more affected by even the slightest of hiccups than any of these activities. Far more complex data sets are being transmitted, and they're needed on a far timelier basis and from all users.

Unlike a video call, which effectively has one creator and several spectators, a virtual world typically comprises many shared participants. Accordingly, loss of any one individual (no matter how temporary) affects the entire collective experience. And even if a user isn't lost altogether, but instead falls slightly out of sync with the rest of the call, they lose their ability to affect the virtual world altogether.

Imagine playing a first-person shooter game. If Player A lags 75 milliseconds behind Player B, they might shoot in a location where they believe Player B to be, but that Player B and the game's server know Player B has already left. This discrepancy means the virtual world's server must decide whose experiences are "true" (that is, which should be rendered and persist across all participants) and whose experiences must be rejected. In most cases the experience of the participant who lagged will be rejected so that the other participants can proceed. The Metaverse can't really function as a parallel plane for human existence if many of those within it experience conflicting (and then invalidated) versions of it.

The computational constraints around the number of users per simulation (which I'll discuss in the next section) often means that if a user disconnects from a given session, they can never rejoin it, either. This disrupts not just that user's experience, but also that of their friends, who must exit the virtual world if they want to resume play together, or otherwise continue without him or her.

In other words, latency and lags might frustrate individual Netflix and Zoom users, but in a virtual world, these problems place the individual at risk of virtual death and the collective in a state of constant frustration. As of this writing, only three-quarters of American households can consistently participate in most real-time rendered virtual worlds. Fewer than one-quarter of households in the Middle East can.

This extended description of the challenge of synchronicity is critical to understanding how the Metaverse will evolve and grow over the coming decades. Although many consider the Metaverse to be reliant upon innovations in devices, such as VR headsets, game engines (such as Unreal), or platforms like *Roblox*, networking capabilities will define—and constrain—much of what's possible, when, and for whom.

As we'll review in later chapters, there are no simple, inexpensive, or quick solutions. We will need new cabling infrastructure, wireless standards, hardware equipment, and potentially even overhauls to foundational elements of the Internet Protocol Suite, such as the Border Gateway Protocol.

Most people have never heard of BGP, but this protocol is everywhere around us, serving as a sort of traffic guard of the digital era by managing how and where data is transmitted across various networks. The challenge with BGP is that it was designed for the internet's original use case of sharing static, asynchronous files. It does not know, let alone understand, what data it's transmitting (be it an email, a live presentation, or a set of inputs intended to dodge virtual gunfire in a real-time rendered virtual simulation), nor its direction (inbound or outbound), the impact of encountering network congestion, and so on. Instead, BGP follows a fairly standardized one-size-fits-all methodology for routing traffic, which essentially weighs the shortest path, the fastest past, and the cheapest path (with a general preference for the last variable). Thus, even if a connection is sustained, it could be an unnecessarily long (latent) one—and could be severed in order to prioritize network traffic that didn't need to be delivered in real time.

BGP is managed by the Internet Engineering Task Force and can be revised. However, the viability of any changes depends on opt-in from thousands of different internet service providers, private networks, router manufacturers, content delivery networks, and more. Even a substantial update is likely to be insufficient for a globally scaled Metaverse—at least in the near future.

## Unlimited Users and Individual Presence

Although Stephenson did not provide an exact date, various references in *Snow Crash* suggest the novel takes place in the mid-to-late 2010s. Stephenson's Metaverse, which was roughly two-and-a-half times the size of earth, was "occupied by twice the population of New York City"[3] at any given time. In total, 120 million of the roughly eight billion people who lived in Stephenson's fictional "real world" had access to computers powerful enough to handle the Metaverse's protocol and could join whenever they liked. In our real world, we are nowhere close to achieving the same.

How far are we? Even nonpersistent virtual worlds that are less than ten square kilometers in surface area, severely constrained in functionality, operated by the most successful video game companies in history, and running on even more powerful computing devices still struggle to sustain more than 50 to 150 users in a shared simulation. What's more, 150 concurrent users (CCUs) is a significant achievement, and only possible because of how these titles are creatively designed. In *Fortnite: Battle Royale*, up to 100 players can participate in a richly animated virtual world, and each player controls a detailed avatar that can use more than a dozen different items, perform dozens of dances and maneuvers, and build complex structures tens of stories tall. However, *Fortnite*'s roughly 5 km² map means that only one dozen to two dozen players will run across one another at once—and by the time players are forced into a smaller portion of the map, most players have been eliminated and turned into data on a scoreboard.

The same technical limitations shape *Fortnite*'s social experiences, such as its famous 2020 concert with Travis Scott. In that case, "players" converged on a much smaller portion of the map, meaning the average device had to render and compute far more information. Accordingly, the title's standard cap of 100 players per instance was halved, while many items and actions, such as building, are disabled, thereby further reducing the workload. While Epic Games can rightly say that more than 12.5 million people attended this live concert, these attendees were split across 250,000 separate copies (meaning, they watched 250,000 versions of Scott) of the event that didn't even start at the same time.

Another good example of the challenges of concurrent users is *World of Warcraft*, a "massively multiplayer online game." To play, users must first pick a "realm"—a discrete server which manages a complete copy of the roughly 1,500-square-kilometer virtual world, and from which they cannot see or interact with any other. In this sense, it may be more accurate to call the game "Worlds" of Warcraft. Users can move between realms, thereby philosophically uniting these many worlds into a single, "massively multiplayer" online game. However, each realm is capped to several hundred participants, and if there are too many users in a specific area, the game creates several distinct and temporary copies of this area, while splitting groups of users among them.

*EVE Online* stands apart from games like *World of Warcraft* and *Fortnite* because all users are part of one singular and persistent realm. But again, this is possible only due to its specific design. For example, the nature of space-based combat also means that action is limited in variety, fairly simple (think laser beams versus jumping or dancing players), and rare. Ordering a ship to mine resources from a planet, or send a succession of blasts from and to a fixed position, is far less complex than a pair of individually animated avatars dancing, jumping, and shooting one another. *EVE Online* is less about what the game processes and renders, but instead what humans plan and decide outside of it. And because the game is set in the vastness of space, most users are far away from one another—enabling CCP Games' servers to effectively treat them as being in separate virtual worlds until necessary. In addition, through the creative use of "travel time," users cannot instantaneously converge on the same location—and there is a strategic cost/risk to leaving a given location.

Even so, *EVE Online* inevitably encounters concurrency problems. At one point in the 2000s, a group of players realized that a specific star system, Yulai, sat near many high-traffic planets inside a major star cluster, making it an enticing spot to establish a new trading hub.[4] They were right. Not long after setting up shop, many buyers began to flock to the area, which attracted additional sellers, then more buyers, and so on. Ultimately, the number of transactions that were occurring inside this hub made the CCP Games' servers start to buckle, leading the publisher to alter the *EVE Online* universe so that the destination would be less convenient to visit.

The lessons from "the Yulai Problem" doubtlessly helped CCP Games design, expand, and overhaul its maps in the years that followed. However, it doesn't help the publisher avoid another outcome: the sudden outbreak of battles so strategically important that thousands of users suddenly converge to save their faction or defeat another.

In January 2021, the largest battle in *EVE*'s history occurred. It was more than twice the size of the prior record and the culmination of a nearly seven-month escalation between the Imperium Faction and a coalition of enemies called PAPI. Or at least, it should have been. The only real losers were CCP Games' servers, which could not keep up with 12,000 players appearing in a single system, and any of those players who were hoping for a decisive victory. Roughly half of the players were unable to ever enter the system, while many of those who did were placed in a sort of purgatory—if they logged into the game, they'd likely be destroyed before having a chance to enter any coherent commands, while leaving meant that their server spot might be taken up by an enemy that would destroy their allies. There was an eventual winner—Imperium—but this was mostly by default, as the defender naturally wins in a battle that never really takes place.

Concurrency is one of the foundational problems for the Metaverse, and for a fundamental reason: it leads to exponential increases in how much data must be processed, rendered, and synchronized per unit of time. It isn't difficult to render an incredibly lush virtual world that no one can touch because it's effectively the same as watching a video of a meticulously designed and predictable Rube Goldberg machine.* And if players—or in this case, viewers—can't affect this simulation, they

---

* These are intricate, chain reaction–styled machines that perform relatively simple tasks through a complex sequence of events. For example, a ball might be placed in a cup by first tipping over domino, which in turn hits many other dominos, ultimately turning on a fan which blows the ball down a rail, before the ball flies into the air, falls down a series of platforms, and then finally lands into its destined cup.

don't need to be continuously connected to or synchronized with it in real time, either.

The Metaverse will only become "the Metaverse" if it can support a large number of users experiencing the same event, at the same time, and in the same place, without making substantial concessions in user functionality, world interactivity, persistence, rendering quality, and so on. Just imagine how different—and limited—society would be today if only 50 to 150 people could attend any given sporting match, concert, political rally, museum, school, or mall.

However, we are far from being able to replicate the density and flexibility of the "real world." And it is likely to remain impossible for some time. During Facebook's 2021 Metaverse keynote, John Carmack, the former and now consulting CTO of Oculus VR (which Facebook bought in 2014 to kickstart its Metaverse transformation) mused that, "If someone had asked me in the year 2000, 'could you build the metaverse if you had one hundred times the processing power you have on your system today . . .' I would have said yes." Yet 21 years later, and with the backing of one of the world's most valuable and Metaverse-focused companies, he believed the Metaverse remained at least five to ten years away and there would be "serious optimization" tradeoffs in realizing this vision—even though there were now billions of computers that were a hundred times more powerful than the hundreds of millions of PCs operating at the turn of the century.[3]

## What's Missing from This Definition

So now we understand my definition of the Metaverse: "A massively scaled and interoperable network of real-time rendered 3D virtual worlds that can be experienced synchronously and persistently by an effectively unlimited number of users with an individual sense of presence, and with continuity of data, such as identity, history, entitlements, objects, communications, and payments." Many readers might be surprised that this definition, as well as its sub-descriptions, are all missing the terms "decentralization," "Web3," and "blockchain." There is good reason for this surprise. In recent years, these three words have become both ubiquitous and entangled—with each other, and with the term "Metaverse."

Web3 refers to a somewhat vaguely defined future version of the internet built around independent developers and users, rather than lumbering aggregator platforms such as Google, Apple, Microsoft, Amazon, and Facebook. It is a more decentralized version of today's internet that many believe is best enabled by (or at least most likely through) blockchains. This is where the first point of conflation begins.

Both the Metaverse and Web3 are "successor states" to the internet as we know it today, but their definitions are quite different. Web3 does not directly require any 3D, real-time rendered, or synchronous experiences, while the Metaverse does not require decentralization, distributed databases, blockchains, or a relative shift of online power or value from platforms to users. To mix the two together is a bit like conflating the rise of democratic republics with industrialization or electrification—one is about societal formation and governance, the other is about technology and its proliferation.

The Metaverse and Web3 may nevertheless arise in tandem. Large technological transitions often lead to societal change because they typically provide a greater voice to individual consumers and enable new companies (and thus individual leaders) to emerge—many of which tap into widespread dissatisfaction with the present to pioneer a different future. It's also true that many companies focused on the Metaverse opportunity today—especially insurgent tech/media start-ups—are building their companies around blockchain technology. As such, the success of these companies would likely lead to a rise in blockchain technology, too.

Regardless, the *principles* of Web3 are likely critical to establishing a thriving Metaverse. Competition is healthy for most economies, and many observers believe that the current mobile generation of the internet and computing is too concentrated among a handful of players. In addition, the Metaverse will not be built directly by the underlying platforms that enable it—just as the US federal government did not build the United States, nor the European Parliament build the

European Union. Instead, it will be constructed by independent users, developers, and small-to-medium businesses, just like the physical world. Anyone who wants the Metaverse to exist—and even those who don't—should want the Metaverse to be driven by (and primarily benefit) these groups rather than by megacorporations.

There are also other Web3 considerations, such as that of trust, which are key to the health and prospects of the Metaverse. Under centralized database and server models, Web3 advocates argue that so-called virtual or digital entitlements are a façade. The virtual hat, plot of land, or movie that a user buys cannot truly be theirs because they cannot ever control it, remove it from the server owned by the company which "sold" it, or ensure that the supposed seller won't delete it, take it back, or alter it. With roughly $100 billion spent on such items in 2021, centralized servers obviously do not prevent considerable user spending; however, it stands to reason this spending is constrained by the need to rely on trillion-dollar platforms which will forever prioritize its interest over those of the individual user. Would you, for example, invest in a vehicle which a dealer might reclaim at any point, or renovate a house which the government might expropriate without cause or redress, or an artwork that the painter might take back once it had appreciated? The answer is sometimes, but definitely not to the same degree. This dynamic is particularly problematic for developers, who must build virtual stores, businesses, and brands despite the inability to guarantee they'll be allowed to operate in the future (and might instead find that the only way to operate is to pay their virtual landlord twice the rent). Legal systems may eventually be updated to provide users and developers with greater authority over their wares, data, and investment, but decentralization, some claim, makes the reliance on court orders unnecessary and their very existence inefficient.

Yet another question is whether centralized server models can ever support a nearly infinite, persistent, world-scale Metaverse. Some believe that the only way to provide the computing resources needed for the Metaverse is through a decentralized network of individually owned—and compensated—servers and devices. But I'm getting ahead of myself.

# Chapter 4

# THE NEXT INTERNET

MY DEFINITION OF THE METAVERSE SHOULD PRO-
vide some insight into why it is often thought of, and fairly described
as, a successor to the mobile internet. The Metaverse will require the
development of new standards and creation of new infrastructure,
potentially require overhauls to the long-standing Internet Protocol
Suite, involve the adoption of novel devices and hardware, and might
even alter the balance of power between technology giants, independent developers, and end users.

The enormity of this transformation also explains why companies
are repositioning themselves in expectation of the Metaverse even
though its arrival remains far off and its effects largely unclear. As
shrewd business leaders know well, every time a new computing and
networking platform emerges, the world and the companies that lead
it are forever changed.

In the mainframe era, which ran from the 1950s through the 1970s,
the dominant computing operating systems were those of "IBM and the
Seven Dwarfs," typically defined as Burroughs, Univac, NCR, RCA,
Control Data, Honeywell, General Electric. The personal computer era,
which began in earnest in the 1980s, was briefly led by IBM and its operating system. However, the eventual winners were new entrants, most
notably Microsoft, whose Windows operating system and Office software suite ran on nearly every PC in the world, and manufacturers such
as Dell, Compaq, and Acer. In 2004, IBM exited the business altogether,
selling its ThinkPad line to Lenovo. The story of the mobile era takes a
similar shape. New platforms rose or emerged, namely those of Apple's

iOS and Google's Android, with Windows falling out of the category altogether and PC-era manufacturers displaced by new entrants such as Xiaomi and Huawei.*

Indeed, generational changes in computing and networking platforms routinely disrupt even the most stagnant and protected categories. In the 1990s, for example, chat services such as AOL Instant Messenger and ICQ quickly established text-based communications platforms which rivaled the customer bases and usage of many telephone companies and even postal services. In the 2000s, these services were surpassed by those focused on live audio, such as Skype, which also connected to traditional and offline phone systems. The mobile era saw a new crop of leaders such as WhatsApp, Snapchat, and Slack. These players didn't just focus on offering Skype but were made for mobile devices. They built services predicated upon different usage behaviors, needs, and even communication styles.

WhatsApp, for example, is intended for nearly constant use—not scheduled or occasional calls, as was the case with Skype—and it is a forum where emojis articulate more than typed words. Whereas Skype was originally built around the ability to make low-to-no-cost calls to the traditional "Public Switched Telephone Network" (i.e., telephones connected to telephone lines), WhatsApp skipped this feature altogether. Snapchat saw mobile communication as being image-first and the front-facing camera on smartphones as more important than the more frequently used (and higher-resolution) back camera, and built numerous AR lenses to enhance that experience as a result. Slack, for its part, built a productivity-based tool for business with programmatic integration into various productivity tools, online services, and more.

Another example comes from the even more regulated and stagnant payments space. In the late 1990s, peer-to-peer digital payment networks such as Confinity and Elon Musk's X.com, which merged to

---

* Another important leader in the mobile device market is Samsung, which, unlike these other manufacturers, is 80 years old. However, it never held significant market share in the mainframe nor PC markets.

form PayPal, rapidly became consumers' preferred method of sending money. By 2010, PayPal was processing nearly $100 billion in payments per year. A decade later, this sum exceeded $1 trillion (in part due to the acquisition of Venmo in 2012).

We can already see precursors to the Metaverse. In platforms and operating systems, the most talked about contenders are virtual world platforms like *Roblox* and *Minecraft*, and real-time rendering engines such as Epic Games' Unreal engine and Unity Technologies' eponymous engine. These all run on an underlying operating system, such as iOS or Windows, but they often intermediate these platforms from developers and end-users. Discord, meanwhile, operates the largest communications platform and social network focused on video gaming and virtual worlds. In 2021 alone, over $16 trillion was settled through blockchain/cryptocurrency networks, which to many experts are foundational enablers of the Metaverse (more on this in Chapter 11). Visa, as a point of contrast, processed an estimated $10.5 trillion.[1]

Understanding the Metaverse as the "next-generation internet" helps explain much more than its potential for disruption. Consider, once again, that there is no plural form of the term "internet." There is no "Facebook internet" or "Google internet." Instead, Facebook and Google operate platforms, services, and hardware that in turn operate on the internet—a literally defined "network of networks"* operating independently, with different technical stacks, but sharing common standards and protocols. There were no strict technical obstacles to a single company developing, then owning and controlling, the Internet Protocol Suite (and some, such as IBM, tried to push their own proprietary suite as part of the so-called Protocol Wars). However, most generally believe this would have led to a smaller, less lucrative, less innovative internet.†

---

\* The term "internet" is an abbreviation of "inter-networking."

† It has been argued that the internet is regionalizing, most notably the Chinese internet, and to a lesser extent, that in the EU. To the extent this claim is valid, it would be due to regulatory enforcement that results in key (and required) differences in standards, services, and content.

We should expect the establishment of the Metaverse to be broadly similar to that of the internet. Many will try to build or co-opt the Metaverse. One of these groups might even succeed, as Sweeney fears. However, it's more likely that the Metaverse will be produced through the partial integration of many competing virtual world platforms and technologies. This process will take time. It will also be imperfect, inexhaustible, and face significant technical limitations as a result. But it is the future we should hope for and work towards.

Moreover, the Metaverse will not replace or fundamentally alter the internet's underlying architecture or protocol suite. Instead, it will evolve to build on top of it in a way that will feel distinctive. Think about the "current state" of the internet. We refer to it as the mobile internet era, yet most internet traffic is still transmitted via fixed-line cables—even for data sent from and to mobile devices—and mostly runs on standards, protocols, and formats designed decades ago (though they've evolved since). We also continue to use some software and hardware designed for the early internet—such as Windows or Microsoft Office—that have evolved since, but are broadly unchanged from decades ago. Despite this, it's clear that the "mobile internet era" is distinct from the predominantly fixed-line internet era of the 1990s and early 2000s. We now primarily use different devices (made by different companies) in new places, for different purposes, using different types of software (mostly apps, rather than general purpose software and web browsers).

We also recognize that the internet is a bundle of many different "things." To interact with the internet, the average person typically uses a web browser or app (software), which they access through a device that can itself connect to "the internet" using various chipsets, all of which communicate using various standards and common protocols, which are transmitted through physical networks. Each of these areas collectively enable internet experiences. No one company could drive end-to-end improvements in the internet—even if they operated the entire Internet Protocol Suite.

# Why Video Games Are Driving the Next Internet

If the Metaverse is indeed a successor to the internet, it might seem odd that its pioneers come from the video gaming industry. After all, the arc of the internet thus far is quite different.

The internet originated in government research labs and universities. Later, it expanded into enterprise, then small-to-medium businesses, and later still, consumers. The entertainment industry was arguably one of the last segments of the global economy to embrace the internet, with the "Streaming Wars" only really beginning in 2019—nearly 25 years after the first public demonstration of streaming video. Even audio, one of the simplest media categories to deliver over IP, remains a mostly non-digital medium, with terrestrial radio, satellite radio, and physical media comprising nearly two-thirds of US recorded music revenues in 2021.

The mobile internet was not led by government, but its arc was broadly the same. When it launched in the early 1990s, usage and software development was concentrated in government and enterprise, then by the late 1990s and early 2000s, small-to-medium businesses. Only after 2008, with the launch of the iPhone 3G, did the mass market adopt it, with consumer-centric apps emerging, for the most part, in the decade that followed.

If we look more closely at this history, we can see why gaming, a $180-billion leisure industry, seems poised to alter the $95-trillion world economy. The key is to consider the role of constraints in all technical development.

When the internet emerged, bandwidth was limited, latency was considerable, and computer memory and processing power scarce. This meant that only small files could be sent and it still took a great deal of time. Almost all consumer use cases, such as photo sharing, video streaming, and rich communications were impossible. But the primary business need—sending messages and basic files (an unformatted Excel sheet, stock purchase orders)—was exactly what the internet was designed to support. The immensity of the service economy, and the importance of managerial functions in the goods economy, was such that even modest productivity enhancements were extraordinarily valuable. Mobile was similar. Early devices couldn't play games or send photos—and streaming a video or FaceTime call was out of the question. However, push email was orders of magnitude more helpful than pager notifications or live phone calls.

Given their complexity, it should be obvious that real-time rendered 3D virtual worlds and simulations were even more constrained by the early decades of the personal computer and internet than almost all other types of software and programs. To this end, governments, enterprises, and small-to-medium businesses had little to no use for graphics-based simulations. A virtual world that can't realistically simulate a fire isn't helpful for firefighters, a bullet that doesn't curve with gravity doesn't aid military snipers, and an architectural firm can't design a building based on the generic idea of "heat from the sun." But video games—*games*—don't need realistic fire, gravity, or thermodynamics. What they need is to be fun. And even an 8-bit, monochromic game can be fun. The consequence of this fact has compounded for nearly 70 years.

For decades, most technically capable CPUs and GPUs owned by a household or small business were typically a video game console or gaming-focused PC. No other computing software required the horsepower of a game. In 2000, Japan even placed export limitations on its own beloved giant, Sony, fearing that the company's new PlayStation 2 device could be used for terrorism on a global scale (for instance, to process missile guidance systems).[2] The following year, in touting the importance of the consumer electronics industry, US Secretary of Commerce Don Evans stated "yesterday's supercomputer is today's PlayStation."[3] In 2010, the US Air Force Research Laboratory built the 33rd largest supercomputer in the world using 1,760 Sony PlayStation 3s. The project's director estimated that the "Condor Cluster" was 5% to 10% the cost of equivalent systems and used 10% of the energy.[4] The supercomputer was used for radar enhancement, pattern recognition, satellite imagery processing, and artificial intelligence research.[5]

The companies that typically focused on powering video game consoles and PCs are now some of the most powerful technology companies in human history. The best example is computing and system-on-a-chip giant Nvidia, which is far from a household name yet ranks alongside consumer-facing tech platforms Google, Apple, Facebook, Amazon, and Microsoft as one of the ten largest public companies in the world.

Nvidia's CEO, Jensen Huang, didn't start his company with the intention of it becoming a gaming giant. In fact, he founded it based on the belief that eventually graphics-based computing would be needed to solve queries and problems that general-purpose computing never could. But to Huang, the best way to develop the necessary capabilities and technologies was to focus on video games. "The condition is extremely rare that a market is simultaneously large and technologically demanding," Huang told *Time* magazine in 2021. "It is usually the case that the markets that require really powerful computers are very small in size, whether it's climate simulation or molecular-dynamics drug discovery. The markets are so small, it [*sic*] can't afford very large investments. That's why you don't see a company that was founded to do climate research. Video games were one of the best strategic decisions we ever made."[6]

Nvidia was founded only a year after *Snow Crash*—which the gaming community also quickly considered to be a seminal text. Despite this, Stephenson has said the emergence of the Metaverse via gaming is "the thing I totally missed in" the novel. "When I was thinking up the Metaverse, I was trying to figure out the market mechanism that would make all of this stuff affordable. *Snow Crash* was written when 3D imaging graphics hardware was outrageously expensive, only for a few research labs. I figured that if it were ever going to become as cheap as TV, then there would have to be a market for 3D graphics as big as the market for TV. So the Metaverse in *Snow Crash* is kind of like TV. . . . What I didn't anticipate, what actually came along to drive down the cost of 3D graphics hardware, was games. And so the virtual reality that we all talked about and that we all imagined 20 years ago didn't happen in the way that we predicted. It happened instead in the form of video games."[7]

For similar reasons, the software solutions that are best at real-time 3D rendering come from gaming, too. The most notable examples are Epic Games' Unreal Engine, as well as Unity Technologies' eponymous engine, but there are dozens of video game developers and publishers with highly capable proprietary real-time rendering solutions.

Non-gaming alternatives exist, but at least for now, they're widely regarded as inferior for real time, specifically because that constraint wasn't necessary to them from the start. The rendering solutions designed for manufacturing or film did not need to process an image in 1/30th or 1/120th of a second. Instead, they prioritized other objectives, such as maximizing visual richness or the ability to use the same file format to both design and manufacture an object. These solutions were typically designed to run on high-end machines, rather than nearly every consumer device worldwide.

One advantage that's often overlooked is the fact that game developers, publishers, and platforms have had to fight and work around the internet's networking architecture for decades and thus have unique expertise as we shift to the Metaverse. Online games have required synchronous and continuous networking connections since the late 1990s, with Xbox, PlayStation, and Steam supporting real-time audio chat across most of their titles since the mid-2000s. Making this work has required predictive AI that takes over for a player during a network drop before handing back control, custom software to unnoticeably "roll back" gameplay in the event that one player suddenly receives information before another, and creating gameplay that aligned with, rather than ignored, the technical challenges likely to affect most players.

This design orientation leads to the final advantage games companies possess: the ability to create a place someone would actually want to spend time in. Daniel Ek, the co-founder and CEO of Spotify, has argued that the dominant business model of the internet era has been breaking down anything made of atoms into bits—what was once a physical alarm clock on a nightstand is now an application inside the smartphone on a nightstand, or just data stored on a smart speaker nearby.[8] In a simplified sense, the Metaverse era can be thought of as involving the use of bits to produce 3D alarm clocks made of virtual atoms. Those with the most experience in virtual atoms—decades of it—are game developers. They know how to make not just a clock, but a room, a building, and a village populated by happy players. If humanity is ever to move to a "massively scaled interoperable network of real-time rendered 3D virtual worlds," that skill is going to take us there. When discussing what he got right and wrong about the future in *Snow Crash*, Stephenson told *Forbes* that "instead of people going to bars on the Street in *Snow Crash*, what we have now is *Warcraft* guilds" which go on in-game raids.[9]

In the first part of this book, I've detailed where the term "Metaverse" and its ideas come from, the various efforts to construct it over the past several decades, as well as its importance to our future. I've investigated the corporate enthusiasm for this would-be successor to the mobile internet, reviewed how this confusion was and continues to be significant, introduced a workable definition that explains what the Metaverse is, and touched on why video game makers seem to be at its forefront. Now, I'll walk you through what it will take to make the Metaverse a reality.

*Part II*

# BUILDING THE METAVERSE

# Chapter 5

# NETWORKING

VERSIONS OF THE THOUGHT EXPERIMENT "IF A tree falls in a forest and no one is around to hear it, does it make a sound?" can be traced back hundreds of years. The exercise endures in part because it's fun, and it's fun because it hinges on important technicalities and philosophical ideas.

The subjective idealist George Berkeley, to whom the above question is often attributed, argued that "to be is to be perceived." The tree—standing, falling, or fallen—exists if someone or something is perceiving it. Others claim that what we mean by "sound" is just vibrations that propagate through matter, and that it exists whether they're received by an observer or not. Or maybe sound is the sensation experienced by a brain when these vibrations interact with nerve endings—and if there are no nerves to interact with vibrating particles, there can be no sound. Then again, for decades humans have been able to produce physical equipment capable of interpreting vibrations as sound, thereby enabling sound to be heard through an artificial observer. But does that count? Meanwhile, the quantum mechanics community today largely agrees that without an observer, existence is at best a conjecture that cannot be proved or disproved—all that can be said is that the tree *might* exist. (Albert Einstein, who was instrumental in founding the theory of quantum mechanics, took issue with this view.)

In Part II, I explain what it will take to power and build the Metaverse, starting with networking and computing capabilities, and then moving on to the game engines and platforms that operate its many virtual worlds, the standards which are needed to unite them, the devices through which they're accessed, and the payment rails that underpin their economies. Throughout these many explanations, I want you to keep in mind Berkeley's tree.

Why? Because even if the Metaverse is "fully realized," it will not *really* exist. It, alongside each of its trees, their many leaves, and the forests they're situated in, will just be data stored in a seemingly infinite network of servers. While it could be argued that as long as that data exists, the Metaverse and its contents do too, there are many different steps and technologies required for it to exist to anyone but a database. Furthermore, each part of the "Metaverse stack" provides a company with leverage and informs what is and isn't possible for another. For example, you'll find out that no more than a few dozen people today can even observe the fall of a high-fidelity tree. And to reach more users? Well, the virtual world will need to be duplicated— in other words, for many people to hear a single tree fall, many trees must fall. (Try that one, Berkeley!) Or, perhaps, its observers are put on time delay, thus also unable to affect the fall nor prove its correlation. Another technique is to simplify the tree's bark to textureless, uniform brown and the sound of its fall into a generic thud.

To unpack these constraints and their implications, I want to begin with a real example: the virtual world I consider the most technically impressive today. No, it's not *Roblox*, nor *Fortnite*. In fact, this virtual world is likely to reach fewer people in its lifetime than each of those titles do in a single day. It's not even fair to call it a game, as so many of the virtual worlds we've looked at so far can be. Instead, it's designed to precisely reproduce an experience many deem unpleasant, dull, or terrifying: airplane travel.

## Bandwidth

The first *Flight Simulator* was released in 1979 and quickly established a small following. Three years later (but still two decades before the first Xbox was released), Microsoft acquired a license for the title, releasing another ten entries by 2006. In 2012, Guinness World Records named *Flight Simulator* the longest-running video game franchise, though it remained unknown to most gamers. It took until the 12th entry, which released in 2020, for *Microsoft Flight Simulator* (*MSFS*) to soar into the public consciousness. *Time* magazine named it one of the best games of the year. The *New York Times* said *MSFS* provided "a new way of understanding the digital world," providing a view "that is more real than the one we can see outside [and] a picture that illuminates our understanding of reality."[1]

In theory, *MSFS* is what many people see it as: a game. Within seconds of opening the application, you'll be reminded that Microsoft's Xbox Game Studios developed and published it. However, the goal of *MSFS* is not to win, kill, shoot, defeat, beat, or score on another player, nor an AI-based competitor. The goal is to a fly a virtual plane—a process that involves much of the same work as flying a real one. Players will communicate with air traffic control and their copilots, wait to be cleared for takeoff, set their altimeter and flaps, check for fuel reserves and mixtures, release their brake, slowly push the throttle, and so on, all before following their selected or designated flight path while managing for conflicting routes and accommodating the flight paths of other virtual planes.

Every entry in the *MSFS* series offered that sort of functionality, but the 2020 edition is extraordinary—the most realistic and expansive consumer-grade simulation in history. Its map is over 500,000,000 square kilometers—just like the "real" planet earth—and includes two trillion uniquely rendered trees (not two trillion copy-and-pasted trees, or two trillion trees made up of a few dozen varieties), 1.5 billion buildings, and nearly every road, mountain, city, and airport across the world.[2] Everything looks like the "real thing" because *MSFS*'s virtual world is based on high-quality scans and images of the "real thing."

*Microsoft Flight Simulator*'s reproductions and renders aren't perfect, but they are still stunning. "Players" can fly past their own house and spot their mailbox or the tire swing in the front yard. Even when the "game" must reproduce a sunset that's reflected across a bay

and that is refracted yet again by the plane's wings, it can be difficult to distinguish between a screenshot from *MSFS* and a real-world
photograph.

To pull this off, *MSFS*'s "virtual world" is nearly 2.5 petabytes
large, or 2,500,000 gigabytes—roughly 1,000 times larger than *Fortnite*. There is no way for a consumer device (or most enterprise devices)
to store this amount of data. Most consoles and PCs top out at 1,000
gigabytes, while the largest consumer-grade Network Attached Storage (NAS) drive is 20,000 gigabytes and retails for close to $750 alone.
Even the physical space required to store 2.5 petabytes is impractical.

But even if a consumer could afford such a hard drive and have
enough space to house it, *MSFS* is a *live* service. It updates to reflect
real-world weather (including accurate wind speed and direction, tem-
perature, humidity, rain, and light), air traffic, and other geographic
changes. This allows a player to fly into real-world hurricanes, or
to trail real-life commercial airliners on their exact flight path while
they're in the air in the real world. This means that users cannot "pre-
buy" or "pre-download" all of *MSFS*—much of it doesn't yet exist!

*Microsoft Flight Simulator* works by storing a relatively small portion of the "game" on a consumer's device—roughly 150 GB. This
portion is enough to run the game—it contains all of the game's code,
visual information for numerous planes, and a number of maps. As a
result, *MSFS* can be used offline. However, offline users see mostly procedurally generated environments and objects, with landmarks such as
Manhattan broadly familiar but populated with generic, mostly duplicated buildings that bear only an occasional and sometimes accidental
resemblance to their real-world counterparts. Some preprogrammed
flight paths exist, but they cannot mimic actual live routes, nor can one
player see another player's plane.

It's when players go online that *MSFS* becomes such a wonder,
with Microsoft's servers streaming new maps, textures, weather data,
flight paths, and whatever other information a user might need. In a
sense, players experience the *MSFS* world exactly as a real-world pilot
might. When they fly over or around a mountain, new information

streams into their retinas via light particles, revealing and then clarifying what's there for the first time. Before then, a pilot knows only that, logically, *something* must be there.

Many gamers assume this is what happens in all online multiplayer video games. But the truth is that most online games try to send as much information as possible to the user in advance, and as little as possible when they're playing. This explains why playing a game, even a relatively small one such as *Super Mario Bros.*, requires purchasing digital discs that contain multi-gigabyte game files, or spending hours downloading these files—and then spending even more time installing them. And then from time to time, we might be told to download and install a multi-gigabyte update before we can play again. These files are so large because they contain nearly the entire game—namely its code, game logic, and all the assets and textures required for the in-game environment (every type of tree, every avatar, every boss battle, every weapon, and so on).

For the typical online game, what actually comes from online multiplayer servers? Not much. *Fortnite*'s PC and console game files are roughly 30 GB in size, but online play involves only 20–50 MB (or 0.02–0.05 GB) in downloaded data per hour. This information tells the player's device what to do with the data they already have. For example, if you're playing an online game of *Mario Kart*, Nintendo's servers will tell your Nintendo Switch which avatars your opponents are using and should therefore be loaded. During the match, your continuous connection to this server enables it to send a constant stream of data on exactly where these opponents are ("positional data"), what they're doing (sending a red shell at you), communications (e.g., your teammate's audio), and various other information, such as how many players are still in the match.

That online games remain "mostly offline" is even a surprise to avid gamers. After all, most music and video is now streamed—we don't predownload songs or TV shows anymore, let alone buy physical CDs to store them—and video gaming is supposedly a more technically sophisticated and forward-looking media category. Yet it is precisely because games are so complicated that those who make them choose to marginalize the internet—because the internet is not reliable. Connections are not reliable, bandwidth is not reliable, latency is not reliable. As I discussed in Chapter 3, most online experiences can survive this unreliability, but games cannot. As a result, developers chose to rely as little on the internet as possible.

This mostly offline approach to online games works well, but it imposes many limitations. For example, the fact that a server can only tell individual users which assets, textures, and models should be rendered means that every asset, texture, and model must be known and stored in advance. By sending rendering data on an as-needed basis, games can have a much greater diversity of virtual objects. *Microsoft Flight Simulator* aspires for every town to not just differ from one another, but to exist as they do in real life. And it doesn't want to store 100 types of clouds and then tell a device which cloud to render and with what coloring; rather, it wants to say exactly what that cloud should look like.

When a player sees their friend in *Fortnite* today, they can interact using only a limited set of pre-loaded animations (or "emotes"), such as a wave or a moonwalk. Many users, however, imagine a future where their live facial and body movements are re-created in a virtual world. To greet a friend, they won't pick Wave 17 of the 20 waves pre-loaded onto their device, but will wave uniquely articulated fingers in a unique way. Users also hope to bring their myriad virtual items and avatars across the myriad virtual worlds connected to the Metaverse. As *MSFS*'s file size suggests, it's simply not possible to send so much data to the user in advance. Doing so not only requires impractically large hard drives, but a virtual world that knows in advance anything that might be created or performed.

The need to "presend" an otherwise living virtual world has other implications. Every time Epic Games alters *Fortnite*'s virtual world—say, to add new destinations, vehicles, or non-playable characters—users have to download and install an update. The more Epic adds, the longer this takes and the longer a user must wait. The more frequently a world is updated, the more delays a user will experience.

The batch-based update process also means that virtual worlds cannot be truly "alive." Instead, a central server is choosing to send a specific version of a virtual world out to all users, a world that will endure until the next update replaces it. Each edition is not necessarily fixed—an update might have programmed changes, such as a New Year's Eve event or snowfall which increases on a daily basis—but it is pre-scripted.

Finally, there are limitations to where users can go. During Travis Scott's 10-minute event on *Fortnite*, some 30 million players were instantly transported from the game's core map to the depths of a never-before-seen ocean, then to a never-before-seen planet, and then deep into outer space. Many of us may imagine that the Metaverse operates in similar fashion—that users can easily jump from virtual world to virtual world without first enduring long loading times. But to put on the concert, Epic had to send users each of these mini-worlds days to hours before the event via a standard *Fortnite* patch (users who hadn't downloaded and installed the update before the event started weren't able to participate in it). Then, during each set piece, every player's device was loading the next set piece in the background. It was notable that each of Scott's concert destinations was smaller and more limited than the one before it, with the last being a largely "on rails" experience in which users simply flew forward in largely nondescript space. Think of this as the difference between freely exploring a mall versus traversing one via a moving walkway.

The concert was nevertheless a significant creative achievement, but as is often the case with online games, it depended on technical choices that cannot support the Metaverse. In fact, the most Metaverse-like virtual worlds today are embracing a hybrid local/cloud streaming data model in which the "core game" is preloaded, but several times as much data are sent on an as needed basis. This approach is less important for titles such as *Mario Kart* or *Call of Duty*, which have relatively small item and environmental diversity, but critical to those like *Roblox* and *MSFS*.

Given the popularity of *Roblox* and the immensity of *MSFS*, it might seem as though modern internet infrastructure can now handle

Metaverse-style live data streaming. However, the model only works today in a highly constrained fashion. *Roblox*, for example, doesn't need to cloud stream much data because most of its in-game items are based on "pre-fabs." The game is mostly just telling a user's device how to tweak, recolor, or re-arrange previously downloaded items. In addition, *Roblox*'s graphical fidelity is relatively modest and therefore its texture and environmental file sizes are relatively small, too. Overall, *Roblox*'s data usage is much greater than that of *Fortnite*—roughly 100–300 MB per hour, rather than 30–50 MB—but still manageable.

At its target settings, *MSFS* needs nearly 25 times as much hourly bandwidth than *Fortnite* and five times as much as *Roblox*. This is because it's not sending data on how to reconfigure or recolor a pre-loaded house, but instead sending a user's device the exact dimensions, density, and coloration of a multi-kilometer cloud or a nearly exact replica of the Gulf of Mexico's shoreline. Yet even this need is simplified in ways that won't work for "the Metaverse."

While *MSFS* needs lots of data, it doesn't need it particularly fast. As with real-world pilots, *MSFS* pilots cannot suddenly teleport from New York State to New Zealand, nor see downtown Albany from 30,000 feet above Manhattan, nor descend from the firmament to the tarmac in a few minutes. This provides the player's device with lots of time to download the data it needs—and even the ability to predict (and thus start downloading) what it needs before the player even selects a destination. Even if this data doesn't arrive on time, the consequences are modest: some of Manhattan's buildings will temporarily be procedurally generated, rather than resemble the real thing, with the realistic details then added when they arrive.

Finally, *MSFS*'s virtual world has more in common with a diorama than Neal Stephenson's bustling and unpredictable Street. Sending users this sort of data, which cannot easily be predicted and is far more voluminous than the visual detail of an office park or forest, will require significantly more than 1 GB per hour. This brings us to the next, and arguably least understood, element of internet connectivity today: latency.

# Latency

Bandwidth and latency are often conflated, and the mistake is understandable: they both impact how much data can be sent or received per unit of time. The classic way to differentiate the two is by likening your internet connection to a highway. You can think of "bandwidth" as the number of lanes on the highway, and "latency" as the speed limit. If a highway has more lanes, it can carry more cars and trucks without congestion. But if the highway's speed limit is low—perhaps due to too many curves or because it's laid in gravel not pavement—then the flow of traffic is slow even if there's spare capacity. Similarly, a high speed limit with only one lane results in constant congestion, too—the speed limit is an aspiration, not a reality.

The challenge with real-time rendered virtual worlds is that users aren't sending a single car from one destination to another. Instead, they're sending a never-ending fleet of cars tethered together (remember, we need a "continuous connection") both to and from that destination. It's not possible to send these cars in advance because their contents are decided only milliseconds before they hit the road. What's more, we need these cars to move at their fastest possible speed and without ever being diverted to another route (which would sever the continuous connection and lengthen the transit time even if the top speed was maintained).

A global road system that meets and sustains these specifications is a substantial challenge. In Part I, I explained that few online services today need ultra-low latency. It doesn't matter if it takes 100 milliseconds or 200 milliseconds or even two-second delays between sending a WhatsApp message and receiving a read receipt. It also doesn't matter if it takes 20 ms or 150 ms or 300 ms after a user clicks YouTube's pause button until the video stops—and most users probably don't register the difference between 20 ms and 50 ms. When you're watching Netflix, it's more important that the video plays reliably rather than immediately. And while latency in a Zoom video call is annoying, it's easy for participants to manage; they just learn to wait a bit after the speaker stops speaking. Even a second (1,000 ms) is workable.

The human threshold for latency is incredibly low in interactive experiences. A user must instinctively feel as though their inputs actually have an effect—and delayed responses mean that the "game" is responding to old decisions after new decisions have been made. For this same reason, playing against a user with lower latency can often feel like you're competing against someone in the future—someone with superspeed—who is able to parry a blow you've not even made.

Think back to the last time you watched a movie or a TV show on a plane, an iPad, or in a theater, and the audio and video were slightly out of sync. The average person doesn't even notice a synchronization issue unless the audio is more than 45 ms early, or over 125 ms late (170 ms total variance). Acceptability thresholds, as they're generally called, are even wider, at 90 ms early and 185 ms late (275 ms). With digital buttons, such as the YouTube pause button, the average person only thinks that their click failed if a response takes 200 to 250 ms. In games such as *Fortnite*, *Roblox*, or *Grand Theft Auto*, avid gamers become frustrated after 50 ms of latency (most game publishers hope for 20 ms). Even casual gamers feel input delay, rather than their inexperience, are to blame at 110 ms.[3] At 150 ms, games that require a quick response are simply unplayable.

So what does latency look like in practice? In the United States, the median time for data to be sent from one city to another and back again is 35 ms. Many pairings exceed this, especially between cities with high-density and intense demand peaks (for example, San Francisco to New York during the evening). Crucially, this is just city-to-city or data center–to–data center transit time. There is still city center–to-the-user transit time, which is particularly prone to slowdowns. Dense cities, local networks, and individual condominiums can easily congest and are often laid with copper cabling with limited bandwidth, rather than high-capacity fiber. Those who live outside a major city might sit at the end of dozens or even hundreds of miles of copper-based transmission. For those whose last mile is on wireless, 4G adds up to 40 additional milliseconds.

Despite these challenges, round-trip delivery times in the United States are typically within the acceptability threshold. However, all connections suffer from "jitter," the packet-to-packet variance in delivery time relative to the median. While most jitter is tightly distributed around the connection's median latency, it can frequently spike severalfold due to unforeseen congestion somewhere along a network path—including the end-users' network as a result of interference from other electronic devices, or perhaps a family member or neighbor initiating a video stream or download. Though temporary, this can easily ruin a fast-paced game or result in a severed network connection. Again, networks aren't reliable.

To manage latency, the online gaming industry has developed a number of partial solutions and workarounds. For example, most high-fidelity multiplayer gaming is "match made" around server regions. By limiting the player roster to those who live in the northeastern United States, or Western Europe, or Southeast Asia, game publishers can minimize latency within each region. Because gaming is a leisure activity and typically played with one to three friends, this clustering works well enough. You're unlikely to want to play with a specific person several time zones away, and you don't really care where your unknown opponents live anyway (in most cases, you can't even talk with them).

Multiplayer online games also use "netcode" solutions to ensure synchronization and consistency and to keep players playing. Delay-based netcode will tell a player's device (say, a PlayStation 5) to artificially delay its rendering of its owner's inputs until the more latent player's (their opponent's) inputs arrive. This will annoy players with muscle memory attuned to low latency, but it works. Rollback netcode is more sophisticated. If an opponent's inputs are delayed, a player's device will proceed based on what it expects to happen. If it turns out the opponent did something different, the device will try to unwind in-process animations and then replay them "correctly."

Although these workarounds are effective, they scale poorly. Net-code works well for titles in which player inputs are fairly predictable, such as driving simulations, or those with relatively few players to synchronize, as is the case in most fighting games. However, it's exponentially harder to correctly predict and coherently synchronize the behaviors of dozens of players, especially when they're participating in a sandbox-style virtual world with cloud-streamed environmental and asset data. That is why Subspace, a real-time bandwidth technology company, estimates that only three-quarters of American broadband homes can consistently (but far from flawlessly) participate in today's high-fidelity real-time virtual worlds, such as *Fortnite* and *Call of Duty,* while in the Middle East less than one-quarter can. And meeting the latency threshold is not sufficient. Subspace has found that an average 10 ms increase or decrease in latency reduces or increases weekly play time by 6%. What's more, this correlation holds beyond the point at which even avid gamers can recognize network latency—if their connection is at 15 ms, rather than 25 ms, they will likely play 6% longer. Almost no other type of business faces such sensitivity, and in that gaming is an engagement-based business, the revenue implications are considerable.

This might seem like a game-specific problem, rather than a Metaverse problem. It's also notable that these issues affect only a portion of game revenues at that. Many hit titles, such *Hearthstone* and *Words with Friends*, are either turn-based or asynchronous, while other synchronous titles, such as *Honour of Kings* and *Candy Crush*, need neither pixel perfect nor millisecond-precise inputs. Yet the Metaverse will require low latency. Slight facial movements are incredibly important to human conversation. We're also highly sensitive to slight mistakes and synchronization issues—which is why we don't mind how the mouths on a cartoonish Pixar character moves, but are quickly creeped out by a photo-real CGI human whose lips don't move exactly right (animators call this the "uncanny valley"). Talking to your mother as though she's on a 100-ms delay can quickly become eerie. While interactions in the Metaverse don't have the time sensitivities of a pixel-specific bullet, the volume of data required is much greater. Recall that latency and bandwidth collectively affect how much information can be sent per unit of time.

Social products, too, depend on how many users can and do use them. Although most multiplayer games are played with other people in the same time zone, or perhaps one removed, internet communication often spans the entire globe. Earlier, I mentioned that it can take 35 ms to send data from the US Northeast to the US Southeast. It takes even longer to travel between continents. Median delivery times from the US Northeast to Northeast Asia are as much as 350 or 400 ms—and even longer from user to user (as much as 700 ms to 1 full second). Just imagine if Face-Time or Facebook didn't work unless your friends or family were within 500 miles. Or they only worked when you were at home. If a company wants to tap into foreign or at-distance labor in the virtual world, it will need better than half-second delays. Every single additional user to a virtual world only compounds synchronization challenges.

Augmented reality–based experiences have particularly stringent requirements for latency because they're based on both head and eye movements. If you wear eyeglasses, you may take it for granted that your eyes immediately adjust to your surroundings when you turn around and receive light particles at 0.00001 ms. But imagine how you'd feel if there were 10–100-ms delays on receiving that new information.

Latency is the greatest networking obstacle on the way to the Metaverse. Part of the issue is that few services and applications need ultra-low-latency delivery today, which in turn makes it harder for any network operator or technology company focused on real-time delivery. The good news here is that as the Metaverse grows, investment in lower latency internet infrastructure will increase. However, the fight to conquer latency doesn't just stretch our pocketbooks; it comes up against the laws of physics. To quote the CEO of one leading video game publisher with experience building games for cloud delivery: "We are in a constant battle with the speed of light. But the speed of light is and will remain undefeated." Consider how difficult it is to send even a single byte from NYC to Tokyo or Mumbai at ultra low-latency levels. At a distance of 11,000–12,500 km, this commute takes light 40–45 ms. The physics of the universe only beat the target minimum for competitive video games by 10%–20%. This doesn't sound like we're losing to the laws of physics. But in practice, we fall far short of this 40–45-ms benchmark. The average latency of a packet sent from Amazon's northeastern US data center (which serves NYC) to its Southeast Asia Pacific (Mumbai and Tokyo) data center is 230 ms.

There are many causes for this delay. One is silica glass. While many assume that data sent via fiber optic cables travel at the speed of light, they're both right and wrong. The light beams themselves do travel at the speed of light—which, as any student knows, is a constant—but they don't travel in a straight line even when the cable itself is laid in a straight line. This is because all glass fibers, unlike the vacuum of space, refract light. As such, the path of a given beam is closer to a tight zigzag bouncing between the edges of a given fiber. The result is a nearly 31% elongation of a route. This brings us to 58 or 65 ms.

In addition, most internet cables are not laid as the crow flies—they must navigate international rights, geographic impediments, and cost/benefit analyses. As a result, many countries and major cities lack a direct connection. NYC has a direct undersea cable to France, but not to Portugal. Traffic from the United States can go directly to Tokyo, but reaching India requires jumping from one undersea cable to another on the Asian or Oceanian continent. A single cable could be laid from the United States to India, but it would need to navigate through or around Thailand—adding hundreds or even thousands of miles—and that only solves shore-to-shore transmission.

Perhaps surprisingly, it's harder to improve domestic internet infrastructure than international internet infrastructure. Laying (or replacing) cables means working around extensive transportation infrastructure (freeways and railways), various population centers

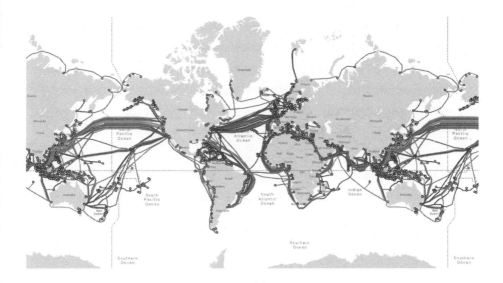

**Image 1. Undersea Cables**
A map of the nearly 500 submarine cables and
1,250 landing stations that enable the global internet.
*TeleGeography*

(each with their own political processes, constituencies, and incentives), and protected parks and lands. Laying a cable over a seamount in international waters is simple compared to laying a cable over a private-public mountain range.

The phrase "internet backbone" might bring to mind a largely planned out, and partly federated, network of cables. In truth, the internet backbone is really a loose federation of private networks. These networks were never laid to be nationally efficient. Rather, they serve local purposes. For example, a private network operator company might have installed a fiber line between two suburbs or even two office parks. Given the expense of permits and the efficiencies of piggybacking onto existing efforts, rather than connect a pair of cities as the crow flies, cable has often been laid when and where other infrastructure was being built.

When data is sent between two cities, such as New York and San Francisco, or Los Angeles and San Francisco for that matter, it may be carried by several different networks strung together (each segment is called a hop). None of these networks were designed to minimize the distance or transit time between these two locations. Accordingly, a given packet might travel substantially farther than the literal geographic distance between a user and server.

This challenge is exacerbated by the Border Gateway Protocol (BGP), one of the core application layer protocols of TCP/IP. As you read in Chapter 3, BGP serves as a sort of air traffic controller for data transmitted "on the internet" by helping each network determine which other network to route data through. However, it does so without knowing what is being sent, in which direction, or with what significance. As such, it "helps" by applying a fairly standardized methodology which mostly prioritizes cost.

BGP's ruleset reflects the internet's original asynchronous network design. Its goal is to ensure all data is transmitted successfully and inexpensively. But as a result many routes are far longer than necessary—and inconsistently so. Two players located in the same building in Manhattan could be in the same *Fortnite* match, managed by a Virginia-based *Fortnite* server, with packets that might be routed through Ohio first and thus take 50% longer to reach the destination. Data might be sent back to one of the players through an even longer network path which runs through Chicago. And any one of these connections might end up severed, or suffer from recurring bouts of 150-ms latency, all to prioritize traffic that didn't need to be delivered in real time, such as an email.

All of these factors together explain why it takes more than four times longer for the average data packet to travel from NYC to Tokyo than a particle of light, five times longer from NYC to Mumbai, and two to four times longer to reach San Francisco, depending on the moment.

Improving delivery time will be incredibly expensive, difficult, and slow. Replacing or upgrading cable-based infrastructure is not only costly—it also requires government approvals, typically at multiple levels. The more direct the intended path of these cables, the more difficult these approvals tend to be because the more direct path is more likely to run into prior residential, commercial, government, or environmentally protected property.

It's much easier to upgrade wireless infrastructure. 5G networks are primarily billed as offering wireless users "ultra-low latency," with the potential of 1 ms and a more realistic 20 ms expected. This represents 20–40 ms in savings versus today's 4G networks. However, this only helps the last few hundred meters of data transmission. Once a wireless user's data hits the tower, it moves to fixed-line backbones.

Starlink, SpaceX's satellite internet company, promises to provide high-bandwidth, low-latency internet service across the United States, and eventually the rest of the world. However, satellite internet doesn't achieve ultra-low latency, especially at great distances. As of 2021, Starlink averages 18–55-ms travel time from your house to the satellite and back, but this time frame extends when the data has to go from New York to Los Angeles and back, as this involves traveling across multiple satellites or traditional terrestrial networks.

In some cases, Starlink even exacerbates the problem of travel distances. New York to Philadelphia is around 100 miles in a straight line and potentially 125 miles by cable, but over 700 miles when traveling to a low-orbit satellite and back down. Not only that, fiber optic cable is much less "lossy" than light transmitted through the atmosphere, especially during cloudy days. Dense city areas, because they are noisy, are subject to interference for that reason, too. In 2020, Elon Musk emphasized that Starlink is focused "on the hardest-to-serve customers that [telecommunications companies] otherwise have trouble reaching."[4] In this sense, satellite delivery enables more people to meet the minimum latency specifications for the Metaverse, rather than offers improvements for those who already meet it.

The Border Gate Protocol may be updated or supplemented with other protocols, or new proprietary standards could be introduced and adopted. In any event, we like to imagine that what's possible is only limited by the minds and innovations of Roblox Corporation, or Epic Games, or the individual creator, and it is true that these groups have proven adept at designing around network-based limitations. They will continue to do so as we navigate all of the bandwidth and latency challenges ahead. At least in the near future, however, these all-too-real limitations will continue to constrain the Metaverse and everything in it.

# Chapter 6

# COMPUTING

SENDING ENOUGH DATA AND IN A TIMELY FASHION is just one part of the process of operating a synchronized virtual world. The data must also be understood, code must be run, inputs assessed, logic performed, environments rendered, and so on. This is the job of central processing units (CPUs) and graphics processing units (GPUs), broadly described as "compute."

Compute is the resource that performs all digital "work." For decades, we've seen increases in the number of computing resources available and manufactured per year, and we've witnessed how powerful they can be. Despite this, computing resources have always been and will likely remain scarce—because when more computing capability is available, we tend to try and perform more complicated calculations. Observe the size of the average video game console over the past 40 years. The first PlayStation, released in 1994, weighed 3.2 pounds and measured 10.75 inches by 7.5 inches by 2.5 inches. The fifth, released in 2020, weighs 9.9 pounds and is 15.4 inches by 10.2 inches by 4.1 inches. Most of the growth relates to the decision to place more computing power in the device—and larger fans to cool it as it performs its work. Today, the original PlayStation (save for its optical drive) could fit in a wallet and cost less than $25, but there's little demand for such a device compared to modern alternatives.

Earlier in the book, I wrote about the supercomputer Pixar built to produce 2013's *Monsters University*: some 2,000 conjoined industrial-grade computers with a combined 24,000 cores. The cost of this data center would have been in the tens of millions of dollars, far more than a PlayStation 3, of course, but also capable of far larger, more detailed, and more beautiful images. Altogether, each of the film's 120,000 frames took 30 core hours to render.* In the following years, Pixar replaced many of these computers and cores with newer and more capable processors that could render these same shots more quickly. But instead of optimizing speed, Pixar uses this power to create more sophisticated renders. For example, one shot in the studio's 2017 film *Coco* had nearly eight million individually rendered lights. At first, it took over 1,000 hours, then 450, to render every frame in the shot. Pixar was able to reduce the time to 55 hours in part by "baking" a number of lights in 20-degree longitudinal and latitudinal increments—that is, reducing their responsiveness to the camera.[1]

This might seem to be an unfair anchor. After all, not every render needs eight million lights, or real-time specifications, nor will it be scrutinized on a 350-square-meter IMAX screen. However, the renders and calculations required for the Metaverse are far more complicated still. They must also be created every ~0.016 or, better yet, ~0.0083 seconds! Not every company—and certainly few individuals—can afford a supercomputer data center. It's actually remarkable how computationally limited even the most impressive virtual worlds are today.

Let's return to *Fortnite* and *Roblox*. While these titles are incredibly creative achievements, their underlying ideas are far from new. For decades, developers have imagined experiences with dozens of live players (if not hundreds or thousands) in a single, shared simulation, as well as virtual environments limited only by the imagination of the individual user. The problem was they were not technically possible.

While virtual worlds with hundreds of even thousands of "concurrent users" (or CCUs) have been possible since the late 1990s, both the virtual worlds and users in them were severely constrained. *EVE Online* does not allow individual players to congregate via avatars. Instead, users direct large and mostly static ships to relocate in space

---

* As a reminder, this is not a literal 30 hours. Instead, it is 30 *core* hours. One core could spend 30 hours rendering, or 30 cores could spend one hour rendering, etc.

or exchange artillery fire. Dozens of *World of Warcraft* avatars can be rendered in the same place, but model detail is limited, the perspective relatively zooms out, and players have limited control over what each avatar can do. If too many players have converged on a single area, the game's server would temporarily "shard" it into concurrently operating but independent copies of that space. Some games even chose to limit real-time rendering to individual players and select in-game AI, with the entire background pre-rendered and thus impossible for players to affect. Engaging in any of these experiences also required a player to buy a dedicated gaming PC, which could run into the thousands of dollars. Even if such a device wasn't strictly necessary, a user likely had to "turn off" or "turn down" the game's rendering capability or halve the frame rate.

It was only by the mid 2010s that millions of consumer-grade devices could manage a game like *Fortnite*—one with dozens of richly animated avatars in a single match, each one capable of a wide range of actions, and interacting in a vivid and tangible world, rather than the cold vastness of space. It was around this same time that enough affordable servers were available that could manage and synchronize the inputs coming from so many devices.

These computational advancements led to extraordinary change in the video gaming industry. Within years, the most popular (and revenue generative) games in the world were those focused on rich UGC and high numbers of concurrent users (*Free Fire, PUBG, Fortnite, Call of Duty: Warzone, Roblox, Minecraft*). In addition, these games quickly expanded into the sorts of media experiences that were previously "IRL Only" (the Travis Scott concert in *Fortnite*, or Lil Nas X's in *Roblox*). The collective result of these new genres and events was enormous growth in the gaming industry. Over the course of an average day in 2021, over 350 million people participated in a battle royale game—just one genre of high CCU game—and billions were able to do so. In 2016, only 350 million people in the world owned the equipment needed to render a rich 3D virtual world. At its peak in 2021, *Roblox* had 225 million monthly users—a figure over a third higher than the lifetime sales of the best-selling console in history, the Play-Station 2, and two-thirds the size of social networks such as Snapchat and Twitter.

As you might be able to guess by now, these games feel so ahead of their time in part because of specific design decisions that allow them to work around current computation constraints. Most battle royales support 100 players, but they also use enormous maps with numerous "points of interest" to scatter players far from one another. This means that while the server needs to track what every player is doing, each player's device doesn't need to render them or track or process their actions. And while players must ultimately converge on a small space—sometimes the size of a dorm room—the very premise of a battle royale means that almost all players have been defeated at that point. And as the map shrinks, it becomes harder to survive. A battle royale player might need to worry about 99 competitors, but their device faces far fewer.

Still, these tricks only go so far. The mobile-only battle royale *Free Fire*, for example, is one of the most popular games in the world. However, most of its players are in Southeast Asia and South America, where most devices are low-to-mid-range Android devices, rather than more powerful iPhones and high-end Androids. As such, *Free Fire*'s battle royale is limited to 50, not 100. Meanwhile, when titles such as *Fortnite* or *Roblox* operate social events in a more confined space, such as a virtual concert venue, they reduce CCUs to 50 or fewer. They also limit what users can do compared to the standard game modes. The ability to build might be turned off, or the number of dance moves reduced from the normal dozen or two to only a single preset option.

If you have a processor that's not as powerful as the average player's, you'll observe that more compromises have to be made. Devices that are a few years old will not load the custom outfits of other players (as they have no gameplay consequence) and instead just represent them as stock characters. For all the marvels of *Microsoft Flight Simu-*

*lator*, fewer than 1% of desktop or laptop Macs and PCs can even run the title on its lowest-fidelity settings. Part of the reason why *MSFS* is possible on those devices is because so little of its world is real beyond its map, weather, and flight paths.

Of course, computing capabilities improve every year. *Roblox* now supports up to 200 players in its relatively lower-fidelity worlds, with up to 700 players possible in beta testing. However, we remain far from the point at which the only constraint is creative. The Metaverse will involve hundreds of thousands participating in a shared simulation and with as many custom virtual items as they like; full motion capture; the ability to richly modify a virtual world (rather than pick from a dozen or so options) with full persistence; and rendering that world not just in 1080p (typically considered "high definition"), but 4K or even 8K. Even the most powerful devices on earth struggle to do this in real time because every single asset, texture, and resolution increase or added frame and player means an additional draw on scarce computing resources.

Nvidia's founder and CEO, Jensen Huang, imagines the next step for immersive simulations as taking us far beyond more realistic-looking explosions or a more animated avatar. Instead, he envisions the application of the "laws of particle physics, of gravity, of electromagnetism, of electromagnetic waves, [including] light and radio waves . . . of pressure and sound."[2]

Whether the Metaverse will require such fidelity to physics is debatable. The important point here is simply that computing power is always scarce specifically because additional computing capabilities lead to important advances. Huang's desire to bring the laws of physics into a virtual world might seem excessive and impractical, but assuming that it is requires predicting and dismissing the innovations that could come from it. Who would have thought that enabling 100-player battle royales would change the world? What is guaranteed is that the availability of and limitations to compute will shape which Metaverse experiences are possible, for whom, when, and where.

## Two Sides of the Same Problem

We know the Metaverse requires more compute, but exactly how much is needed remains unclear. In Chapter 3, I quoted Oculus's former and now consulting CTO, John Carmack, who believes "building the Metaverse is a moral imperative." In October 2021, Carmack said that if he'd been asked 20 years earlier whether "one hundred times the processing power" would be sufficient to meet this duty, he would have said yes. Yet even though billions of devices now hold such capability, according to Carmack the Metaverse remains at least five to ten years away and would still face "serious optimization tradeoffs" even at the further edge of that prediction. Two months later, Raja Koduri, Intel's SVP and general manager of its Accelerated Computing and Graphics Group published similar thoughts on Intel's investor relations site. Koduri said that "indeed, the metaverse may be the next major platform in computing after the world wide web and mobile . . . [but] truly persistent and immersive computing, at scale and accessible by billions of humans in real time, will require even more: a 1,000-times increase in computational efficiency from today's state of the art."[3]

There are varying perspectives on how best to achieve this.

One argument is that as much "work" as possible should be performed in remote, industrial-grade data centers rather than in consumer devices. That most of the work involved in a virtual world happens on each user's device strikes many as wasteful because it means many devices are performing the same work at the same time in support of the same experience. In contrast, the super-powerful server operated by the virtual world's "owner" is just tracking user inputs, relaying them when necessary, then refereeing process conflicts when they occur. It doesn't even need to render anything!

An example helps bring this to (virtual) life. When a player shoots a rocket launcher at a tree in *Fortnite*, this information (the item used, its attributes, and the trajectory of the projectile) is sent from that player's device to *Fortnite*'s multiplayer server, which then relays that information to all of the players who require that information. Their local machines then process and act on that information: they show the explosion, determine whether their players are harmed, remove the tree from the map, and allow the players to move through where it once was, and so on.

In practice, players might not even see the same visual explosion, even though the "same" explosive hit the exact "same" tree at the exact "same" angle at the exact "same" time, and the exact same logic was applied to process the cause and effect. This reflects the fact that (due to variable latency) a given device might think the rocket was sent slightly earlier or later, and from a slightly different position. Usually this doesn't matter, but sometimes it is enormously consequential. For example, Player 1's console might determine that Player 2 was killed by the explosion that destroyed the tree, while Player 2's console would say Player 2 took significant, but not fatal, damage. Neither console is "wrong," but the game obviously can't proceed with both versions of the "truth." And so the server must "pick."

The current reliance on personal devices creates other limitations, too. Consumers can experience only what their own device can manage. A 2019 iPad, 2013 era PlayStation 4, and 2020 edition PlayStation 5 will all present *Fortnite* differently. The iPad will be limited to 30 frames per second, while the PlayStation 4 will offer 60 FPS and the PlayStation 5 120 FPS. The iPad will likely load only selective map textures and maybe even skip avatar outfits, while the PlayStation 5 will show refracting light and shadows, something the PlayStation 4 cannot. This, in turn, means that the overall complexity of a virtual world ends up partly limited by the lowest end device that can access it. Epic Games has decided that the avatars and outfits in *Fortnite* shouldn't have an impact on its gameplay, but changing its mind might entail cutting off many players.

Shifting as much processing and rendering to industrial-grade data centers seems both more efficient and essential to building the Metaverse. There are already companies and services pointing in this direction. Google Stadia and Amazon Luna, for example, process all video gameplay in remote data centers, then push the entire rendered experience to a user's device as a video stream. The only thing a client device needs to do is play this video and send inputs (move left, press X, and so on)—similar to watching Netflix.

Proponents of this approach often highlight the logic of powering our homes via power grids and industrial power plants, not private generators. The cloud-based model allows consumers to stop buying consumer-grade, infrequently upgraded, and retailer-marked-up computers and instead rent access to enterprise-grade equipment that is more cost-efficient per unit of processing power and more easily updated. Whether a user has a $1,500 iPhone or an old WiFi-enabled fridge with a video screen, they could play a computationally intensive title such as *Cyberpunk 2077* in all its fully rendered glory. Why should a virtual world depend on a small piece of consumer hardware wrapped in plastic dye covers, rather than on a multi-million-dollar (if not billion-dollar) server stack owned by the company that operates the virtual world?

For all the ostensible logic of this approach, and the success of server-side content services such as Netflix and Spotify, remote rendering is not the consensus solution among game publishers today. Tim Sweeney has argued that "initiatives to place real-time processing on the wrong side of the latency wall have always been doomed to failure because, even though bandwidth and latency are improving, local computing performance is improving faster."[4] Put differently, the debate is not whether remote data centers can offer better experiences than consumer-owned ones. They obviously can. Rather, it's that networks get in the way and will likely continue to do so.

Here the power generator analogy begins to break down. In most of the developed world, consumers don't struggle to receive the power they need on a daily basis, nor as quickly as needed. This is despite the fact that very little power—that is, data—is sent. For remote-rendered experiences to be delivered, many gigabytes per hour will need to be sent in real time. But as you know, we're still struggling to send a few megabytes per hour on a timely basis.

Furthermore, remote compute has yet to prove itself to be more efficient for rendering. This is a consequence of several interconnected issues.

First, a GPU does not render an entire virtual world, nor even much of it, at any given point. Instead, it renders just what's necessary for a given user when that user needs it. When a player turns around in a game like *The Legend of Zelda: Breath of the Wild*, the Nintendo Switch's Nvidia GPU effectively unloads everything that was previously rendered in order to support the player's new field of view. This process is called "viewing-frustrum culling." Other techniques include "occlusion," in which objects that are in a player's field of view are not loaded/rendered if they are obstructed by another object, and "level of detail" (LOD), in which information, such as the nuanced texture of a birch tree's bark, are only rendered when the player should be able to see it.

Culling, occlusion, and LOD solutions are essential to real time rendered experiences because they enable a user's device to concentrate its processing power on what the user can see. But as a result, other users cannot "piggyback" off the work of one player's GPU. Some readers might think this is a lie, recalling many hours spent playing *Mario Kart* on the Nintendo 64, which allowed players to "split" a TV screen into four, one for each driver. Even today, *Fortnite* allows a single PlayStation or Xbox to cleave a screen in half so that two players might play at once. But in this case, the relevant GPU is supporting simultaneous renders for multiple participants, not users. The distinction here is critical. Every player must enter the same match and level—and cannot leave it early, either. This is because the device's processors can only load and manage a finite amount of information, and its random-access memory system will temporarily store various renders (e.g., a tree or building) so that it can be continuously reused by each player, rather than rendered from scratch each time. Furthermore, each player's resolution and/or frame rate drops by an amount proportional to the number of users. This means that even if two TVs are used to operate two-player *Mario Kart*, rather than one TV split in two, each player receives half as many rendered pixels per second.*

It is technically possible for a GPU to render two entirely different games. A top-of-the-line Nvidia GPU can certainly support two distinct emulations of a 2D sidescrolling *Super Mario Bros.*, or one version of *Super Mario Bros.* and another similarly low-powered title. However, this is not done in a compute-efficient manner. An Nvidia GPU that might run high-end Game A at its fullest rendering specifications cannot run two versions of the title at half of the specifications— or even a third. It also cannot trade off its power between each game based on what they need and when, like a parent helping two kids study or get to bed. Even if Game A can never use all of a given Nvidia GPU's power, that spare cannot be assigned elsewhere.

GPUs do not generate generic rendering "power" that can be divided across users in the way a power plant splits electricity across multiple homes, or in the way a CPU server can support input, location, and synchronization data for a hundred players in a battle royale. Instead, GPUs typically operate as a "locked instance" supporting a single player's rendering. Many companies are working on this problem, but until it's possible, there's no inherent efficiency in designing "mega GPUs" akin to large industrial power generators, turbines, or other infrastructure. While power generators are typically more cost-efficient per unit of power as their capacity increases, the reverse is true with GPUs. A GPU that's twice as powerful as another, in a simplified sense, costs more than twice as much to produce.

The difficulties of "splitting" or "sharing" GPUs are why Microsoft Xbox's cloud game streaming server farms are, in fact, made up of racks

---

* The exception here is when a game is running well below the capacity of the GPU that's supporting it—as would be the case if one played the Nintendo 64 version of *Mario Kart* on the Nintendo Switch, which released 21 years after the Nintendo 64.

and racks of de-shelled Xboxes, each one serving a player. Put another way, Microsoft's electrical power plant is really just a network of single-household power generators, rather than a single, neighborhood-sized one. Microsoft could use bespoke GPU and CPU hardware to support cloud instances, rather than the GPU and CPU hardware in its consumer-centric Xboxes. However, this this would require every Xbox game be developed to support an additional "type" of Xbox.

Cloud-rendering servers also face utilization issues. A cloud gaming service might require 75,000 dedicated servers for the Cleveland area at 8 p.m. Sunday night, but only 20,000 on average, and 4,000 at 4 a.m. Monday. When consumers own these servers, in the form of consoles or gaming PCs, it doesn't matter that they're unused or offline. However, data-center economics are oriented toward optimizing for demand. As a result, it will always be expensive to rent high-end GPUs with low utilization rates.

This is why Amazon Web Services gives customers a reduced rate if they rent servers from Amazon in advance ("reserved instances"). Customers are guaranteed access for the next year because they've paid for the server, and Amazon is pocketing the difference between the cost and the price the customer is charged (AWS's cheapest Linux GPU reserved instance, equivalent to a PS4, costs over $2,000 for one year). If a customer wants to access servers when they need them ("spot instances"), they might find they're not available, or that only lower-end GPUs are. This last point is key: we're not solving the computing shortage if the only way to make remote servers affordable is to use rather than replace older ones.

There is another way to improve cost models: consolidate servers into fewer locations. Rather than operate a cloud game streaming center in Ohio, Washington State, Illinois, and New York, a company could just build one or two. As the number and diversity of customers increases, demand tends to stabilize, resulting in greater average utilization rates. Of course, this also means increasing the distance between remote GPUs and the end user, thereby increasing latency. And this doesn't solve for the distance between users, either.

Shifting computing resources into the cloud creates many new costs. For example, side-by-side, always turned-on devices running at data centers create considerable heat—far more than the aggregate heat of those servers sitting in a family's living room credenza. Servicing, securing, and managing this equipment is costly. The shift from streaming limited bits of data to high-resolution, high-frame-rate content means substantially higher bandwidth costs, too. Yes, Netflix and others make the costs work, but they're typically sending fewer than 30 frames of video per second (not 60 to 120) with a lower resolution (e.g., 1K or 2K, not 4K or 8K, as Google Stadia was promised), on a non-real-time basis, and from nearby servers that are storing files rather than performing intensive computing operations.

For the foreseeable future, what I call "Sweeney's Law"—improvements in local compute will continue to outpace improvements in network bandwidth, latency, and reliability—seems likely to hold. Although many believe that Moore's Law, which was coined in 1965 and states that the number of transistors in a dense integrated circuit doubles about every two years, is now slowing down, CPU and GPU processing power continues to grow at a rapid pace. In addition, consumers today frequently replace their primary computing device, resulting in enormous improvements for end-user compute every two to three years.

## Dreams of Decentralized Computing

The insatiable need for more processing power—ideally, located as close as possible to the user but, at the very least, in nearby industrial server farms—invariably leads to a third option: decentralized computing. With so many powerful and often inactive devices in the homes and hands of consumers, near other homes and hands, it feels inevitable that we'd develop systems to share in their mostly idle processing power.

Culturally, at least, the idea of collectively shared but privately owned infrastructure is already well understood. Anyone who installs solar panels at their home can sell excess power to their local grid (and, indirectly, to their neighbor). Elon Musk touts a future in which your Tesla earns you rent as a self-driving car when you're not using it yourself—better than just being parked in your garage for 99% of its life.

As early as the 1990s programs emerged for distributed computing using everyday consumer hardware. One of the most famous examples is the University of California, Berkeley's SETI@HOME, wherein consumers would volunteer use of their home computers to power the search for alien life. Sweeney has highlighted that one of the items on his "to-do list" for the first-person shooter *Unreal Tournament 1*, which shipped in 1998, was "to enable game servers to talk to each other so we can just have an unbounded number of players in a single game session." Nearly 20 years later, however, Sweeney admitted that goal "seems to still be on our wish list."[5]

Although the technology to split GPUs and share non–data center CPUs is nascent, some believe that blockchains provide both the technological mechanism for decentralized computing as well as its economic model. The idea is that owners of underutilized CPUs and GPUs would be "paid" in some cryptocurrency for the use of their processing capabilities. There might even be a live auction for access to these resources, either those with "jobs" bidding for access or those with capacity bidding on jobs.

Could such a marketplace provide some of the massive amounts of processing capacity that will be required by the Metaverse?* Imagine, as you navigate immersive spaces, your account continuously bidding out the necessary computing tasks to mobile devices held but unused by people near you, perhaps people walking down the street next to

---

* Neal Stephenson described this sort of technology and experience at length in *Cryptonomicon*, which was published in 1999, seven years after *Snow Crash*.

you, to render or animate the experiences you encounter. Later, when you're not using your own devices, you would be earning tokens as they return the favor (more on this in Chapter 11). Proponents of this crypto-exchange concept see it as an inevitable feature of all future microchips. Every computer, no matter how small, would be designed to be auctioning off any spare cycles at all times. Billions of dynamically arrayed processors will power the deep compute cycles of even the largest industrial customers and provide the ultimate and infinite computing mesh that enables the Metaverse. Perhaps the only way for everyone to hear a tree fall is for all of us to water it.

# Chapter 7

# VIRTUAL WORLD ENGINES

A VIRTUAL TREE FALLS IN A VIRTUAL FOREST. IN the previous two chapters, I explained what is required for the tree to be rendered, and for its fall to be processed and then shared and thus known to any observer. But what is this tree? Where is this tree? What is the forest? The answer is data and code.

Data describes the attributes of a virtual object, such as its dimensions or color. For our tree to be processed by a CPU and rendered by a GPU, this data needs to be run by code. And if we want to chop down that tree and use its wood to build a bed or light a fire, that code must be part of a much *broader* framework of code* that operates the virtual world.

The real world isn't altogether different. The laws of physics are the code that reads and runs all interactions—from the reasons a tree falls to how this produces vibrations in the air that travel to a human ear causing nerves to relay information via electrical signal across various synapses. Similarly, a tree "seen" by a human observer means it reflected light produced by (usually) the sun, light that is in turn received and processed by the human eye and brain.

But there is a key difference: the real world is fully preprogrammed. We can't see X-rays or echolocate, but the requisite information already exists in the world. In a game, X-rays and echolocation require data and a lot of code. If you go home, mix ketchup and petroleum, and

---

* The tree may itself be code that brings together many smaller virtual objects, such as leaves, trunks, branches, and bark.

then try to eat it or paint with it, the laws of physics take care of the results. For a game to manage the same interaction, it needs to know in advance what ketchup and petroleum do when they're combined (and probably in generic ratios), or it needs to know enough about the two for the game's logic to figure it out, assuming the game is capable.

A virtual world's logic might say that petroleum cannot be mixed with anything. Or that it can only be mixed with oil. Or that if it's mixed with anything at all, it produces unusable sludge. But a more complicated result requires considerably more data and for the virtual world's logic to be far more comprehensive. How much petroleum can be added to ketchup before it's inedible? How much ketchup can be added to petroleum before it's unusable? How does the resulting substance's color and viscosity change based on the ratio of one ingredient to the other?

The fact that so many of these permutations have little value is, in fact, hugely valuable to those who produce virtual worlds. Because the hero of *The Legend of Zelda* does not need to go to space, no space-based physics are required. *Call of Duty*'s players don't need kayaks, enchantments, or baking goods; the game's developer did not create the relevant code. Nintendo and Activision could focus more data and code on what their virtual worlds need and benefit from, rather than infinite permutations that have limited practical value to their games.

For all its efficiency, this approach introduces obstacles to building Metaverse-like virtual worlds, and especially in establishing interoperability among them. In *Microsoft Flight Simulator*, for example, a pilot can land a helicopter beside a football field, but there's no football game he can then watch, let alone join. For Microsoft to offer such functionality, it would need to build from scratch its own football system, even though many developers have already done so and, with years of experience, are likely better at it, too. While *MSFS* could instead try to integrate into these football-specific virtual worlds, the data structures and codes on every side are likely incompatible. In the networking and compute chapters above, I discussed the fact that users' devices are often performing the same work. But if a comparison can be made, developers must be said to be even worse. They are constantly building and rebuilding everything from a football field to a football and even rules for how a football flies in the air. What's more, this work is getting harder every year as virtual world builders look to take advantage of more sophisticated CPUs and GPUs. According to Nexon, one of the world's largest video game publishers, the average number of staff credits for an open world action game (think *The Legend of Zelda* or *Assassin's Creed*) has grown from roughly 1,000 in 2007 to over 4,000 by 2018, with budgets growing by a factor of 10 (roughly two and half times faster).[1]

To hear trees fall, to have them fall near football fields, and to have the sound of their falling added to the roars of a crowd responding to a game-winning touchdown requires lots of programmers writing lots of code for handling vast amounts of data and all in the same ways.

Now that we've covered the networking and computing power required to share, run, and render the data and code required for the Metaverse, we can turn to these latter concepts.

## Game Engines

The concept, history, and future of the Metaverse are all intimately tied to gaming, as we've seen, and this fact is perhaps most obvious when we look at the basic code of virtual worlds. This code is typically contained in a "game engine," a loosely defined term that refers to the bundle of technologies and frameworks that help to build a game, render it, process its logic, and manage its memory. In a simplified sense, think of the game engine as the thing that establishes the virtual laws of the universe—the ruleset that defines all interaction and possibilities.

Historically, all game-makers built and maintained their own game engines. But the past fifteen years have witnessed the rise of an alternative: licensing an engine from Epic Games, which makes the Unreal

Engine, or one from Unity Technologies, which makes an eponymous engine.

Using these engines has a cost. Unity, for example, charges every individual developer that uses it an annual fee. This fee ranges from $400 to $4,000, depending on the features required and the size of the developer's company. Unreal typically charges 5% of net revenues. Fees aren't the only reason to build your own engine. Some developers believe doing so for a given game genre or experience, such as realistic and fast-paced first-person shooters, ensures that their games "feel better" or perform better. Others worry about the need to rely on another company's pipelines and priorities, or fret that their vendor has such a detailed view into their game and its performance. Given the concerns, it's common for large publishers to build and maintain their own engines (some, such as Activision and Square Enix, even operate a half dozen or more).

Most developers, however, see strong net positives to licensing and then customizing Unreal or Unity. Licensing allows a small or inexperienced team to build a game with an engine more powerful and extensively tested than they could ever build—and that is less likely to fail and can never run over budget. In addition, they can focus more of their time on what will differentiate their virtual world—level design, character design, gameplay, etc.—instead of the basic technology required for it to run. And rather than hiring a developer and then training them to use or build on a proprietary engine, they can instead appeal to the millions of individual developers already familiar with Unity or Unreal and immediately put them to work. For similar reasons, it's also easier to integrate third-party tools. An independent start-up that makes, say, face-tracking software for video game avatars doesn't design their solution to work with a proprietary engine they've never used, but instead to work with those chosen by the largest number of developers.

A good analogy is designing and building a house. Neither the architect nor the decorator designs proprietary lumber dimensions, assembly hardware, measurement systems, blueprint frameworks, or tools. Not only does this make it easier to focus on the creative work, but it makes it easier to hire carpenters, electricians, and plumbers. If the house ever requires a renovation, another team can more easily modify the existing structure because they don't need to learn new techniques, tools, or systems.

This analogy has a key shortfall, however. Houses are built once and in one place. Games, conversely, are designed to run on as many possible devices and operating systems as possible—some of which haven't yet been developed, let alone released. As a result, games must be compatible with, say, different voltage standards (e.g., the UK's 240 volts and America's 120 volts), measurement systems (imperial and metric), conventions (aerial telephony wires and buried ones), and so on. Unity and Unreal build and maintain their game engines so that they're not just compatible with but optimized for every platform.*

In a sense, we can think of independent game engines as a shared R&D pool for the industry. Yes, Epic and Unity are for-profit companies, but instead of every developer sinking part of their budget into proprietary systems to manage core game logic, a few cross-platform technology providers can concentrate a portion of their budgets into a more capable engine that supports, and benefits, the entire ecosystem.

As the major game engines developed, another type of independent gaming solution emerged: live services suites. Companies such as PlayFab (now owned by Microsoft's Azure) and GameSparks (Amazon) operate much of what a virtual world needs to "run" online and multiplayer experiences. This includes user account systems, player data storage, processing in-game transactions, version management, player-to-player communications, matchmaking, leaderboards, game analytics, anti-cheat systems, and more, all of which work across platforms. Both Unity and Epic now have their own live services offerings, too, which are available at low-to-no cost and are not limited to their engines. Steam, the world's largest PC game store and a key

---

* As you'll recall from the discussion of GPUs and CPUs, the fact that Unreal or Unity is compatible with most gaming platforms does not necessarily mean that a given experience can run on them.

point of discussion in Chapter 10, offers its own live services product, Steamworks.

As the global economy continues to shift to virtual worlds, these cross-platform and cross-developer technologies will become a core part of global society. In particular, the next wave of virtual world builders—not game-makers, but retailers, schools, sports teams, construction companies, and cities—are likely to use these solutions. Companies like Unity, Unreal, PlayFab, and GameSparks are in an enviable position. Most obviously, they become a sort of standard feature, or lingua franca, for the virtual world—think of them as the "English" or "metric" of the Metaverse. Just as it is likely that you use some English and some knowledge of the metric system when traveling internationally, odds are that if you're building something online today, irrespective of what it is you're building, you are using—and paying—one or more of these companies.

But more importantly, who better to establish common data structures and coding conventions across virtual worlds than the companies that govern their logic? Who better to facilitate exchanges of information, virtual goods, and currencies between these virtual worlds than the companies that power the same inside them? And who better to create an interconnected network of these virtual worlds, as ICANN does for web domains and IP addresses? We'll come back to these questions and this presumptive answer, but first we need to consider a path some think of as the easier, and best, way to build the Metaverse.

## Integrated Virtual World Platforms

As both independent game engines and live services suites developed over the past two decades, other companies combined these approaches into a new one: integrated virtual world platforms (IVWPs) such as *Roblox*, *Minecraft*, and *Fortnite Creative*.

IVWPs are based around their own general-purpose and cross-platform game engines, similar to Unity and Unreal (*Fortnite Creative*, or FNC, which is owned by Epic Games, is built using Epic's Unreal Engine). However, they are designed so that no actual "coding" is required. Instead, games, experiences, and virtual worlds are built using graphical interfaces, symbols, and objectives. Think of it as the difference between using the text-based MS-DOS and visual iOS, or designing a website in HTML versus creating one in Squarespace. The IVWP interface enables users to create more easily and with fewer people, less investment, and less expertise and skill. Most *Roblox* creators, for example, are kids, and nearly 10 million users have created virtual worlds on *Roblox*'s platform.

In addition, every virtual world built on these platforms must use the platform's entire live services suite—its account and communication systems, avatar database, virtual currency, and more. All of these virtual worlds must be accessed through the IVWP, which therefore serves as a unified experiential layer and a single installer file. In this sense, building a world on *Roblox* is more like constructing a Facebook page than a Squarespace website. *Roblox* even operates an integrated marketplace for developers where they can upload anything they custom-made for their virtual world (e.g., a Christmas tree, a snowed-on tree, a barren tree, a pine bark texture) and license it to other game makers. This provides developers with a second source of income (developer-to-developer rather than just developer-to-player), while also making it easier, cheaper, and faster for others to build their virtual worlds. The process also drives further standardization of virtual objects and data.

Though it's easier for a developer to build a virtual world using an IVWP than a game engine like Unreal or Unity, it's harder to build an IVWP than a game engine in the first place. Why? Because for an IVWP, everything is a priority. An IVWP wants to enable creators' creative flexibility while also standardizing underlying technologies, maximizing interconnectivity among everything that's built, and minimizing the need for training or programming knowledge on the part of creators. Imagine if IKEA wanted to build a country as dynamic as the United States, but force all buildings to use IKEA prefabs. Additionally, IKEA would be in charge of the new country's currency, utilities, police, customs, and more.

A good way to understand how hard it is to operate an IVWP was provided to me by Ebbe Altberg, the former CEO of *Second Life*. In the mid-2010s, one of the platform's developers created a business selling virtual horses, alongside a subscription to virtual horse feed. Later, *Second Life* upgraded its physics engines, but a bug resulted in horses sliding past their feed whenever they tried to eat it. As a result, the horses starved and died. It took time for *Second Life* to even know this bug existed, and more time still to fix it, then to provide the appropriate redress to those affected by it. Still, such events disrupt *Second Life*'s economy, while also producing distrust in the market, which harms both buyers and sellers. Finding a way to constantly improve functionality, while continuing to support old programming and without errors, is an extraordinary task. Game engines also face a version of this problem. However, when Epic updates Unreal, it is up to every developer to deploy this update, and they can do so at the time of their choosing, after extensive testing, and without worrying about how that update affects their interactions with other developers. When *Roblox* pushes an update, it automatically reaches all of its worlds.

At the same time, the fact that a "virtual IKEA" is built on programming, not particle board, means that its potential is bound not by literal physics, but the nearly boundless potential of software. Anything made in *Roblox*, by the Roblox Corporation or its developers, can be endlessly repurposed or copied at no marginal costs. They can even be improved upon. Every developer in an IVWP is effectively collaborating to populate an ever-expanding and increasingly capable network of virtual worlds and objects. As this network improves, it becomes easier to attract more users and more per-user spending, which leads to more network revenue, and then more developers and investment, and thus further improvements to the network, and so on. This is the benefit of pooling not just engine R&D, but, well, R&D for everything.

But what does this look like in practice? The Roblox Corporation offers the best answer at the moment, given that *Fortnite Creative* is managed by Epic Games, which remains private, and *Minecraft*'s financials are not disclosed by its owner, Microsoft.

Start with engagement. By January 2022, *Roblox* was averaging more than 4 billion hours of usage per month, up from roughly 2.75 billion a year earlier, 1.5 billion the year before that, and 1 billion at the end of 2018. This excludes time spent watching *Roblox* content on YouTube, which is the world's most used video site and reports that gaming content is its most watched content category, and *Roblox* its second most popular game (*Minecraft*, another IVWP, ranks first). As a point of contrast, Netflix is estimated at 12.5 billion to 15 billion hours of use per month. All of the top *Roblox* games, such as *Adopt Me!*, *Tower of Hell*, and *Meep City*, come from independent developers with little to no prior experience and staffs of 10 to 30 (having started with one or two). To date, these titles have been played 15 to 30 billion times each. In a single day, they'll reach half as many players as *Fortnite* or *Call of Duty*—and half as many as titles like *The Legend of Zelda: Breath of the Wild* or *The Last of Us* do in their *lifetimes*. And as for populating the platform with a wide range of virtual objects? 25 million items were made in 2021 alone, with 5.8 billion being earned or bought.[2]

Part of *Roblox*'s surging engagement is driven by its growing userbase. From Q4 2018 to January 2022, average monthly players increased from an estimated 76 million to more than 226 million (or 200%), while average daily players grew from around 13.7 to 54.7 (or 300%). You'll note that daily players grew more than the monthly userbase, and engagement grew by an even larger volume (400%). Not only is *Roblox* becoming more popular overall; it's becoming more popular with its users, too. We can see similar evidence of *Roblox*'s network effects in its financials. *Roblox*'s revenues are up 469% from Q4 2018 to Q4 2021, while its payments to on-platform world builders (i.e., developers) have grown 660%. In other words, the average *Roblox* user is spending more per hour than ever before and generating revenues faster than ever before, and with growth in these two metrics exceeding the already impressive growth in users, which is then exceeded by the growth in compensation to developers. Furthermore, *Roblox*'s growth has been disproportionately concentrated among older audiences. At the end of 2018, 60% of daily users were under 13. Three years later, only 21% were. Put another way, *Roblox* ended 2021 with nearly two and half times as many players over 13 as the service had under 13 in 2018.

The most impressive aspect of Roblox Corporation's flywheel may be its investments in R&D. In the first quarter of 2020, the last before the COVID-19 pandemic, the company generated roughly $162 million in revenue and invested $49.4 million in R&D. That means 30 cents of every dollar spent on *Roblox* went back into the platform. Over the following seven quarters, Roblox's revenue surged more than 250%, totaling $568 million in the fourth quarter of 2021. However, Roblox did not divert this revenue to profits, nor any alternative uses. Instead, it continued to reinvest in R&D—at roughly the same rate as before. As a result, the company spent more on R&D in Q4 2021 than *it generated in revenue* in Q1 2020. In 2022, Roblox R&D may top $750 million and by the end of the year, it may approach $1 billion on an annualized basis.

As points of contrast, consider Rockstar's *Grand Theft Auto V* and *Red Dead Redemption 2*. *GTA:V* is the second-best-selling game in history, with over 150 million copies sold (*Minecraft* ranks first with nearly 250 million). *RDR2* was the best-selling title made for the eighth generation of consoles (i.e., PlayStation 4, Xbox One, Nintendo Switch), with 40 million copies sold. The two games are also believed to be among the most expensive game productions ever, with final budgets estimated at $250 million to $300 million and $400 million to $500 million, respectively, which includes more than half a decade of development each, plus extensive marketing and publishing costs. Or compare Roblox's R&D budget to that of Sony's PlayStation group, which topped $1.25 billion in 2021 and spanned close to a dozen gaming studios, its cloud gaming division, live services group, and hardware division. That same year, Epic Games' Unreal Engine is believed to have generated less than $150 million in revenue. Unity's engine brought in much more—roughly $325 million—but still came in 20% short of Roblox's R&D.

Roblox's R&D investments are diverse, spanning improvements in developer tools and software, server architecture to synchronize high concurrency simulations, machine learning to detect harassment, artificial intelligence, rendering for virtual reality, motion capture, and more. That Roblox can invest so much into its platform is astonishing. In theory, every additional dollar enables developers to produce more compelling virtual worlds, which attracts more users leading to more revenue—which enables not just more R&D by Roblox, but also by the independent developers who make these worlds, investment that again drives more user engagement and spending on *Roblox*, leading to more R&D by the company.

## Many Virtual Platforms and Engines, Not Many Metaverses

Think back to the definition of the Metaverse I laid out in Chapter 3: "A massively scaled and interoperable network of real-time rendered 3D virtual worlds that can be experienced synchronously and persistently by an effectively unlimited number of users with an individual sense of presence, and with continuity of data, such as identity, history, entitlements, objects, communications, and payments. Some might read this definition and think *Roblox* is pretty close. It cannot be experienced synchronously and persistently by an effectively unlimited number of users; no real-time rendered virtual world can, at the moment. And when that does become possible, it will surely be true for *Roblox*. However, *Roblox* is unlikely to meet my definition in one key way: most virtual works will exist outside of it. This makes it a Metagalaxy, rather than the Metaverse.

But could *Roblox* become the Metaverse? What if Epic's IVWP *Fort-*

*nite Creative*, game engine Unreal, and live services suite Epic Online Services, along with its other special projects, were combined—would the result be the Metaverse? If you squint, you might be able to imagine these companies, or one like it, subsuming all virtual experiences, thereby becoming a Metaverse-sized Metagalaxy. And it is notable that some form of this process is what happens in *Snow Crash* and *Ready Player One*.

The current state of technological progress, however, suggests another outcome. Why? Because as fast as these virtual giants are growing, the number of virtual experiences, innovators, technologies, opportunities, and developers are all growing faster.

While *Roblox* and *Minecraft* are among the most popular games in the world, their reach is modest when considered in the broadest terms. These two supposed titans have 30–55 million daily active users, a fraction of the global internet population of 4.5–5 billion. In effect, they are still at the ICQ stage of virtual words; billions of users and millions of developers have yet to even try them. It's easy to assume that *Roblox* or *Minecraft* will be the primary beneficiaries of this growth, yet history cautions us to be skeptical.

When Microsoft acquired *Minecraft* developer Mojang in 2014, the title had sold more copies than any other game in history, and also had more monthly active users—25 million—than any AAA video game in history. Seven years later, *Minecraft* had grown nearly five times in monthly users, but also had ceded its crown to *Roblox*, which had grown from fewer than 5 million monthly users to over 200 million. Furthermore, the new king boasts nearly twice the *daily* users as *Minecraft* had *monthly*. What's more, this period included the launch of many other IVWPs. *Fortnite* didn't launch until 2017, with *FNC* coming a year later. Another battle royale, *Free Fire*, which also counts more than 100 million daily active users globally, released its creative mode in 2021. Although it launched in 2013, *Grand Theft Auto V* spent much of the past decade transforming from a single-player game into a makeshift IVWP in *Grand Theft Auto Online*. Sometime over the next few years, the title's much-anticipated next sequel will release and will doubtlessly take advantage of the successes and learnings from *Roblox*, *Minecraft*, and *FNC*.

As long as there are billions, or even tens of millions, of players left to adopt IVWPs, more will come to market. Krafton, one of South Korea's largest companies and the creator of *PUBG*, the first and most popular mainstream battle royale, is surely working on its own offering. In 2020, Riot Games, which makes the most successful game in China, *League of Legends*, bought Hypixel Studios, which previously operated the largest private *Minecraft* server before shutting down to develop their own *Minecraft*-like platform.

Many new IVWPs are being developed around different technical premises, too. At the end of 2021, even the largest of the blockchain-based IVWPs, which includes *Decentraland*, *The Sandbox*, *Crypto-voxels*, *Somnium Space*, and *Upland*, had less than 1% of *Roblox's* and *Minecraft's* daily active users. However, these platforms believe that by allowing users more ownership over their in-world items, as well as a say in how the platform is governed, and a right to share in its profitability, they will be able to grow far more quickly than traditional IVWPs (more on this theory in Chapter 11).

Facebook's *Horizon Worlds* is not limited to immersive VR and AR, but it is focused in those areas, which contrasts with *Roblox*, which is available in immersive VR but prioritizes traditional screen interfaces, such as an iPad or PC screen. Upstarts such as *Rec Room* and *VRChat* are also centered on immersive VR world creation, and are rapidly accumulating users. With valuations around $1 billion–$3 billion each at the end of the 2021, the two platforms remain small. But at the start of 2020, Unity Technology and Roblox Corporation were valued at less than $10 billion and $4.2 billion respectively. Two years later, both have valuations that exceed $50 billion. Niantic, the maker of *Snap* and *Pokémon Go*, is working on its own augmented reality and location-based virtual world platforms.

These competitors might falter, but it's more likely that they grow alongside and potentially displace current market leaders. Take Facebook, as an example. The social networking giant entered 2010 with more than half a billion monthly active users, but has failed to sub-sume any of the hit social media platforms which emerged in the decade. Snapchat launched in 2011, with Facebook launching its own Snapchat-like app (or "clone") in 2013, called "Poke," which was shut-tered a year later. In 2016, Facebook launched "Lifestage," its second Snapchat clone, with was also closed after 12 months. That same year, Facebook's Instagram app also copied Snapchat's signature "Stories" format, with Facebook's main app adding the feature the following year. Then in 2019, Instagram launched its own dedicated Snapchat-like app, "Threads from Instagram," though almost no one noticed. Facebook Gaming, the company's Twitch competitor, launched in 2018, as did Facebook's TikTok competitor, Lasso. Facebook Dating released in 2019, with Instagram adding a TikTok-like feature named "Reels" in 2020. Facebook's efforts have undoubtedly curbed these services' growth, yet each service is larger than ever and still expand-ing. By the end of 2021, TikTok had more than billion users and was reportedly the most visited web domain of the year, with Google and Facebook rounding out the top three.

Though the top integrated virtual world platforms are mighty and fast-growing, they also represent a far smaller portion of the gaming industry than Facebook does in the social web. In 2021, the combined revenues of *Roblox*, *Minecraft*, and *FNC* represented less than 2.5% of gaming revenues in 2021, and reached fewer than 500 million of an estimated 2.5 billion–3 billion players. Moreover, they're dwarfed by the major cross-platform engines. Roughly half of all games today run on Unity, while Unreal Engine's share of high-fidelity 3D immersive worlds is estimated at between 15% and 25%. Roblox's R&D expen-ditures may exceed that of both Unreal and Unity, but this ignores the billions in additional investment made by licensors of these engines. The two most popular games in the world, excluding low-fidelity casual titles such as *Candy Crush*, are *PUBG Mobile* and *Free Fire*, both built on Unity. Most important may be the reach of Unreal and Unity's developers. While millions of users have made a *Minecraft* mod or a *Roblox* game, the number of professional developers using these IVWPs is estimated in the tens of thousands. Epic and Unity count *millions* of active and skilled developers. And scores of proprietary engines, such as Activision's IW (*Call of Duty*) and Sony's Decima (*Horizon Zero Dawn* and *Death Stranding*) continue to receive investment and the games that use them are more popular than ever.

The growing value of virtual worlds and the Metaverse increases the incentives for a developer to in-source their technology stack, as this approach provides a greater opportunity for technical differentiation and greater control over their technology overall, reduces their reliance on third parties who might become competitors,* and increases profit margins. Of course, many of these developers will still use Unreal or Unity as a game engine, or GameSparks or PlayFab for live services. However, these providers enable a developer to "pick and choose" what they like, and also to customize much of what they license. Unlike IVWPs, they also allow a developer to manage its own account systems and operate its own in-game economies. These services are also much cheaper, too. Roblox pays a developer less than 25% of the revenue a player spends on their game.† Epic's Unreal Engine, conversely, takes only a 5% royalty on revenues. The total cost of Unity's engine is likely to be less than 1% of a successful game's

---

* Epic Games' history with *Fortnite* is a good example of this concern. As the highest-grossing game in the world from 2017 through 2020, *Fortnite* has obviously cannibalized players, player hours, and player spend from other games—some of which are made by publishers other than Epic but which used Epic's Unreal Engine. In addition, the version of *Fortnite* which is so popular today—its "battle royale"— was not the original version of the game. When the title launched in July 2017, it was a cooperative survival game in which players worked to defeat zombie hordes. It wasn't until September 2017 that Epic added its battle royale mode, which closely resembled that used by the hit game *PUBG*, which, notably, licensed the Unreal Engine. The publisher behind *PUBG* subsequently sued Epic for copyright infringement, though the suit was later dropped (it's unclear if a settlement was made). In 2020, Epic launched its own publishing arm to release games made by independent studios, thereby placing the company in even greater competition with some of the publishers which occasionally licensed Unreal.

† There is some flexibility here—and most analysts expect this payout ratio to go up over time. More on this topic in Chapter 10.

revenue. Roblox does take on additional expenses for its developers, such as costly server fees, customer service, and billing, but in most cases a developer will still have higher profit potential by building a standalone virtual world, rather than one inside an IVWP. As such, we should assume that no matter how much more successful *Roblox* or *Minecraft* become, they will power only a minority share of all games.

While games and game engines are central to the Metaverse, they don't come close to encompassing it. Most other categories have their own rendering and simulation software. Pixar, for example, builds its animated worlds and characters using its proprietary Renderman solutions. Most of Hollywood, meanwhile, uses Autodesk's Maya software. Autodesk's AutoCAD, along with Dassault Systèmes' CATIA and SolidWorks, are the primary solutions used to build and design virtual objects that will then be made into real-world ones. Examples include cars, buildings, and fighter jets.

In recent years, Unity and Unreal have made inroads into nongaming categories, including engineering, filmmaking, and computer-aided design. In 2019, as discussed earlier, the Hong Kong International Airport used Unity to build a "digital twin" that could be connected to myriad sensors and cameras throughout the airport to track and evaluate passenger flows, maintenance, and more—all in real time. The use of "game engines" to power such simulations does make it easier to produce a Metaverse which spans both the physical and virtual planes of existence. However, the success of the Hong Kong airport endeavor and other simulations like it means more competition, as Autodesk, Dassault, and others respond by adding their own simulation functionality. And just as Unreal and Unity don't provide all the technology required to build or operate a game, they're not sufficient in other domains either. Many new software companies are emerging that take the "stock" editions of these engines and "productize" them for civil and industrial architects, engineers, and facility managers, while also adding their own custom code and functionality. One example is Disney's Industrial Light & Magic (ILM) special effects division. Since using Unity to film Disney's *The Lion King* (2017) and Unreal for the first season of the TV show *The Mandalorian* (2019), ILM has developed its own real-time rendering engine, Helios. The fact that even the most avid *Star Wars* fans failed to notice any impact from the switch from Unreal to Helios for *The Mandalorian*'s second season further suggests how many different rendering solutions and platforms will be built in the years to come.

As measured by the number of assets created, the fastest-growing category of virtual software may be those that scan the real world. Matterport, for example, is a multi-billion-dollar platform company whose software converts scans from devices such as iPhones to produce rich 3D models of building interiors. Today, the company's software is primarily used by property owners to create vivid and navigable replicas of their real estate on sites such as Zillow, Redfin, or Compass, affording would-be renters, as well as construction professionals and other services providers, a better way to understand the space than allowed by blueprints, photographs, or even live tours. Soon we might use such scans to determine the placement of a wireless router or plant, test out a selection of different lamps (each one purchasable through Matterport), or to operate our entire smart home, including electricity, security, HVAC, and more.

Another example is Planet Labs, which scans nearly the entire earth via satellite each day and across eight spectra bands, capturing not just high-resolution imagery, but details including heat, biomass, and haze. The company's goal is to make the entire planet, in all its nuances, legible to software and to update its data on a daily to hourly basis.

Given the pace of change, level of technical difficulty, and the diversity of potential applications, it's likely that we will end up with dozens of popular virtual worlds and virtual world platforms, with many more underlying technology providers. This is a good thing, to my mind. We should not want a single virtual world platform or engine operating the entire Metaverse.

Recall Tim Sweeney's warning about the scope of the Metaverse: "This Metaverse is going to be far more pervasive and powerful than anything else. If one central company gains control of this, they will become more powerful than any government and be a God on Earth."

It's easy to find such a statement hyperbolic, and it may be. Yet we already worry about how the big five technology companies—Google, Apple, Microsoft, Amazon, and Facebook, each one valued in the trillions—manage our digital lives, influencing how we think, what we buy, and more. And right now, most of our lives are still offline. While hundreds of millions of people today are hired through the internet, and work using their iPhones, they don't literally perform their work inside iOS or by building iOS content. When your daughter attends school via Zoom, she accesses Zoom and her school through her iPad or Mac, but the school isn't operated inside the iOS platform. In the West, e-commerce's share of addressable retail spend now hovers between 20% and 30%, but most of this spend is for physical goods, and retail is just 6% of the economy. What happens when we shift to the Metaverse? What happens when a corporation operates the physics, real estate, customs policies, currency, and government of a second plane of human existence? Sweeney's warning starts to sound less hyperbolic.

From a purely technological perspective, we shouldn't want the evolution of the Metaverse to be tied to the investments and beliefs of a single platform. The company Sweeney is imagining would surely prioritize its control over the Metaverse rather than what's best for its economies, developers, or users. It would surely maximize for its share of the profits, too.

But if we don't have a single Metaverse platform or operator—and if we don't want one, either—then we need to find a way to interoperate between them. Here we return, once again, to trees. As you'll see, I wasn't kidding when I said the existence of a virtual tree is harder to ascertain than that of a real one.

# Chapter 8

# INTEROPERABILITY

METAVERSE THEORISTS LIKE TO USE THE TERM "interoperable assets," but this is a misnomer because virtual assets don't exist. Only data does. And it's here—at the very start—where the problems of interoperability begin.

Consider the "interoperability" of physical goods, such as a pair of shoes. The manager of an Adidas store in the "real world" could decide to prohibit a customer from wearing Nikes in their store. This would be a business decision, and obviously a poor one, and almost impossible to enforce. A customer wearing Nikes can get inside an Adidas store by opening its door. This is because physics is universal and thus atoms are "write once, run everywhere." The fact that Nike shoes physically exist means they are automatically compatible inside an Adidas store. The Adidas store manager would need to *create* a system to block non-Adidas shoes, write a policy, and then enforce it.

Virtual atoms do not work in this way. For virtual goods from a virtual Nike store to be understood in a virtual Adidas outlet, the latter would need to admit information on these shoes from Nike, operate a system that understood this information, and then run code to operate the shoes accordingly. Suddenly, the admission of sneakers has changed from passive to active.

Today, there are hundreds of different file formats used to structure and store data. There are dozens of popular real-time rendering engines, most of which have been fragmented further through various code customizations.* As a result, almost all virtual worlds and software systems are incapable of understanding what each considers a "shoe" (data), let alone being able to use that understanding (code).

That such enormous variation could exist might surprise those familiar with common file formats such as JPEG or MP3, or who know most websites use HTML. But the standardization of online languages and media stems from how late "for-profit" businesses came to the internet. iTunes, for example, didn't release until 2001, nearly 20 years after the Internet Protocol Suite was established. It was impractical for Apple to reject the standards already in wide use, such as WAV and MP3. Gaming is a different story. When the industry began to emerge in the 1950s, no standards for virtual objects, rendering, or engines existed. In many cases, the companies producing these games were pioneering computer-based content. Apple's Audio Interchange File Format (AIFF) is still the most common audio file format used to store sound on Apple computers; it was created in 1988 and based on game maker Electronic Arts' 1985 general purpose Interchange File Format standard. Furthermore, video games were never intended to be part of a "network" like the internet. Instead, they existed to run on a fixed and offline piece of software.

Virtual worlds today have so much technical diversity for this reason, but also because of the intense computational and networking demands of modern gaming—everything is purpose-built and individually optimized. AR and VR experiences, 2D and 3D games, realistic and cartoon-styled worlds, high concurrent users and low concurrent user simulations, high-budget and low-budget titles, and 3D printers— all use different formats and store data differently. Full standardization would likely mean underserving one application, falling massively short on another, and so on—often in unpredictable ways.

---

* In Unity, the y-axis in an x/y/z coordinate system for a virtual object refers to up/down, while Unreal uses the z-axis for up/down and maps the y-axis to left/right. Converting this information is easy for software, but the disagreements on such foundational data conventions help us to understand how different conventions are between engines.

**Image 2. From the web comic *xkcd*.**
*xkcd.com*

The challenge goes beyond file formats and approaches more ontological questions. It's relatively easy to agree on what an image is—they're only two dimensional and don't move (with video files just being successions of images). But in 3D, especially with interactive objects, agreement is far harder. For example, is a shoe an object, or is it a collection of objects? And if so, how many? Are the caps on a shoelace part of the shoelace, or separate from it? Does a shoe have a dozen individual eyelets, each of which can be customized or even removed, or are they a single interconnected set? If shoes seem hard, just imagine avatars—would-be representations of real people. Forget trees; what is a person?

Beyond visuals, there are other attributes that must be examined, such as motion or "rigging." The bodies of the Incredible Hulk and a jellyfish should not move in the same way, but this means the creator of these avatars needs to enshrine them with code detailing this movement and which another platform can understand. To permit third-party objects, platforms will also need data that describes a good's appropriateness (e.g., nudity, penchant for violence, language style and tone). A game for toddlers needs to differentiate between a PG-rated swimsuit and an R-rated one. Similarly, a gritty war simulator will want to know the difference between a sniper wearing a tree-like ghillie suit and a sniper that's actually an anthropomorphic tree. All of this requires data conventions, and probably additional systems, too. A 2D game will want to be able to import a 3D avatar, but restyle it accordingly. And vice versa.

So we will need technical standards, conventions, and systems for an interoperable Metaverse. But that's not enough. Think about what happens when you send a send a photo from your iCloud storage to your grandmother's Gmail account—suddenly, your iCloud and her Gmail both have a copy of that image. Your email service does too. And if she downloads it from her email, there are now four copies. Yet this doesn't work for virtual goods if they're to hold value and be traded. Otherwise infinite copies will exist every time they're shared between one world and another, or one user and another. This means that systems are needed to track, validate, and modify ownership rights to these virtual goods, while also safely sharing this data from partner to partner.

If a player buys an outfit in Activision Blizzard's *Call of Duty* and wants to use it in EA's *Battlefield*, how is that to work? Does Activision send the outfit's ownership record to EA, which manages it until it's needed elsewhere, or does Activision indefinitely manage the outfit and provide EA temporary rights to use it? And how is Activision paid to do this? If the player sells the outfit to an EA user who doesn't have an Activision account, what happens then? Which company even processes the transaction? What if the users decide to modify the outfit in the EA game? How is that record altered? If users have virtual items scattered across multiple titles, how do they ever know what, altogether, they own and where what they own can or cannot be used?

The 3D standards to use (or not use), the systems to build and data to structure, the partnerships that need to be struck, the valuable data that must be protected but also shared—these and other issues have real-world financial implications. The largest of these considerations, however, might be how to manage an economy of interoperable virtual objects.

Video games are not designed to "maximize GDP." They're designed to be fun. While many games have virtual economies that enable users to buy, sell, trade, or earn virtual goods, this functionality exists in support of play and as part of the publisher's revenue model. As a result, these publishers tend to manage in-game economies by fixing prices and exchange rates, limiting what can be sold or traded, and almost never allowing users to "cash out" into real-world currency.

Open economies, unrestrained trading, and interoperation into third-party titles all make creating a sustainable "game" much more difficult. The promise of profit naturally brings about work-like incentives for players, but these can erode fun—the game's very purpose. And a level playing field for competition, also part of what makes playing a game fun, can be easily undermined by an ability to buy items that otherwise had to be earned. As many publishers monetize their games by selling in-game cosmetics and objects, they fear the moment their players stop buying their virtual items because they've bought them from a competing developer and then imported them. Given all of this, it's understandable that many publishers would rather focus on making their games better, more appealing, and more popular rather than on connecting into a not-yet-formed virtual goods marketplace with unclear financial value and likely involving technical concessions.

To achieve even a measure of interoperability, the gaming industry will need to align on a handful of so-called interchange solutions— various common standards, working conventions, "systems of systems," and "frameworks of frameworks" that can safely pass, interpret, and contextualize information from or to third parties, and consent to unprecedented (but secure and legal) data-sharing models that allow competitors to both "read" and "write" against their databases and even withdraw valuable items and virtual currency.

## Interoperability Is a Spectrum

Reading about the difficulty of getting many virtual worlds to agree on a tree, or a pair of shoes, or the means of walking up to a tree to cut it down and sell it as a Christmas tree three virtual worlds over, you may be asking whether we can reasonably expect a meaningfully interoperable Metaverse to exist at any future point. The answer is yes, but it requires nuance.

Most apparel is interoperable in the real world. All belts, for example, are presumed to work with all pants. Exceptions exist, of course, but overall, most belts are compatible with most pants, irrespective of the year you bought the belt, the brand of belt, or what country you were in when you bought it. At the same time, not all belts fit all pants equally well. There are common standards for pants and belts, but a 30 × 30 pant from J.Crew fits differently from a 30 × 30 pant from Old Navy (dresses vary even more; European and American shoe size standards are entirely different; and so on).

Globally, many differing technical standards exist, such as those for residential voltage, and for measurements of speed, distance, or weight. In some instances, new equipment is required for a foreign device to be used (for example, an electrical outlet adapter), and in other instances a local government will require replacements, such as a car's exhaust to be replaced to meet local emissions regulations.

Pants work everywhere, though not every location you wish to visit will admit jeans. Movie theaters allow almost any clothes and most forms of credit, but you can't bring in outside food nor drink. One can carry a shotgun in much, but not all, of the American outdoors, but rarely in cities and almost never in a school. Cars work on all roads in the US, but to drive on a golf course you need to rent a golf cart (even if you own one). Not every business accepts every currency, but currencies can be exchanged for a fee. Many stores support some, but not all, credit cards and a few accept none. Most of the world now embraces trade, but not all of it, nor for all things, in all quantities, or for free.

Identity is even more complicated. We have passports, credit scores, school records, legal records, employer IDs, state IDs, and more. Which of them are used for what, which of them are available to outside parties or can be affected by outside parties, all varies—sometimes based on where a person is at a given time.

The internet is not much different. There are still public and private networks (and even offline ones), as well as networks, platforms, and software that admit the majority of common file formats, but not all. While the most popular protocols are free and open, many are paid and private.

Interoperability in the Metaverse is not binary. It is not about whether virtual worlds will or won't share. It's about how many share, how much is shared, when, where, and at what costs. So why am I optimistic that, given all these complications, there will be a Metaverse? Economics.

Start with the question of user spending. Many Metaverse skeptics pose some version of the question, "Who wants to wear *Fortnite*'s Peely skin while playing *Call of Duty*?" Now to be fair, a giant, comically styled anthropomorphic banana doesn't make much sense in *Call of Duty*, or in a virtual classroom, for that matter. But it is equally obvious that some users want some items, such as a Darth Vader costume, a Lakers jersey, or a Prada purse, in many different spaces. And they certainly don't want to buy these items over and over and over again. They might be reluctantly willing to do so today, but that's because we're still in the early stages of the shift to virtual apparel. In 2026, hundreds of millions of people will be sitting on numerous (effectively) duplicated outfits across their many previously played games—and will doubtlessly resist buying those outfits again. Liberating purchases from a single title will, the theory suggests, lead to both more purchases and higher prices.

Put another way, would Disney sell more or less merchandise if it could be worn or used only in its theme parks? How much would someone pay for a Real Madrid jersey that could only be used in Santiago Bernabéu Stadium? Or how much lower would user spending be on *Roblox* if a player's outfit was limited to a single *Roblox* game?

It's likely that consumer spending today is constrained by the very knowledge that no game lasts forever. Think of anything you might buy on holiday but don't plan to bring home in your suitcase—a boogie board, a stainless steel water bottle, a costume for Día de los Muertos. Expected obsolescence always constrains our spending.

The utility of these goods is further limited by ownership restrictions. Most games and gaming platforms prohibit users from giving outfits or items to other users, or even selling them for in-game currency. The publishers that do allow reselling and trading typically place firm limits on this activity. Roblox Corporation only allows "limited items" to be resold (otherwise peer-to-peer trading would undermine the sale of goods from Roblox's own shop)—and only *Roblox Premium* subscribers can sell these items.

What is more, although we might believe we have "purchased" these items, we've really just licensed them and the company can "repossess" them at any time. This isn't a huge problem for $10 skins and dances, but no one will buy $10,000 worth of virtual property that could be taken away from them at any moment, with or without repayment.

Consider a case from early 2021, as reported by *South China Morning Post*'s Josh Ye. Tencent, China's largest gaming company, "sued a game item trading platform to determine who owns the in-game currency and items." Specifically, the company argued that these assets had "no material value in real life" and that in-game coins bought with real money were "effectively services charges."[1] The result was outrage, with many gamers feeling mistreated and/or demeaned.

Ownership rights are foundational to investment and the price of any good, while the opportunity for profit is a well-established motivator. Speculation has always financed the growth of new industries, even when it results in bubbles (a lot of America's now cheap-to-use fiber optic cabling was laid in the run-up to the dotcom crash). If we want the greatest possible investment of time, energy, and money into the Metaverse—if we want to achieve the Metaverse—we need to establish firm ownership rights.

Every stakeholder in virtual worlds faces incentives and risks that point in this direction. It's dangerous for any developer to build a business whose wares or services are limited by the popularity of a given platform or its economy (or economic policies). And anything that results in less investment and thus fewer and worse products overall doesn't benefit the developer, the user, or the game and its platform.

Limiting the range of identity and player data is another impediment to the Metaverse economy. Toxicity in gaming is a significant concern for many, and rightfully so. Today, however, while Activision might ban a player from *Call of Duty* for abusive or racist language, that player can then go on to troll on Epic Games' *Fortnite* (or on Twitter, or Facebook). The player could also just create a new PlayStation Network Account, or change to Xbox Live, and while that means fragmenting his or her achievements, some of these achievements are locked to a given platform anyway. Of course, publishers don't want to make their competitors' games better, nor are they usually inclined to share their play data. But no gaming company benefits from toxic behavior, and everyone is negatively affected by it.

Economics, then, will drive standardization and interoperation over time.

The Protocol Wars offer an illustrative example. From the 1970s to the 1990s, few believed that the many competing networking stacks would be replaced by a single suite, least of all one steered by nonprofit and informal working bodies. Instead, we'd contend with a "cyberspace divided."

Banks and other financial institutions didn't use to share credit data, either—it was considered far too valuable and privileged. But eventually, they were convinced that credit scores with better data and more coverage would be of collective benefit. Competing homestay marketplaces Airbnb and Vrbo are now partnering with a third party to prevent guests with a history of poor behavior from making future bookings. Although this harms the individual offenders, all other guests, hosts, and platforms benefit.

The best example of "economic gravity" comes from the game engines—the very companies pioneering the plumbing of the Metaverse.

**Network Gateways: A Cyberspace Divided.** In this map, the major computer networks huddle in the mass of the Matrix, the term for the global collection of computer networks that can exchange electronic mail. The Internet serves as a common ground for much communication online, with commercial online services building gateways for electronic mail as well as other communication and data protocols to the Internet. Major national services such as France's Minitel (http://www.minitel.fr/) now provide gateway communication from their services to the Internet.

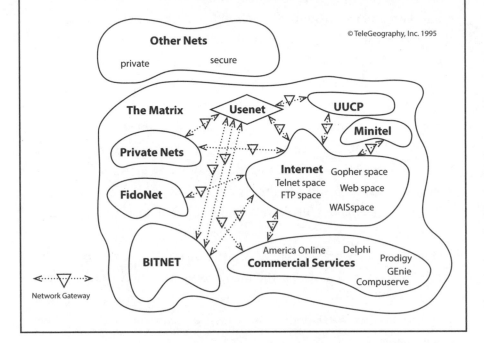

### Image 3. Telegeography Map

This 1995 map and its caption reflect what many experts at the time considered to be the future of online networking: fragmented networks and protocol suites. The internet, in this case, would not be a unifying internetworking standard, but more of a common ground for different collections of networks, some of which would be unable to directly communicate with one another. Most of these networks would exist in "The Matrix," though some would forever sit outside of it. But this future never happened. Instead, the internet became the core gateway between all private and public networks, thereby enabling each network to communicate with any other.

*TeleGeography*

Although the opportunity in virtual worlds has never been as large as it is today, reaching the entirety of this market has never been so difficult. In the 1980s, a developer might make a game for just one console and in doing so, reach 70% of potential players. Two developers might reach all players. Today, there are three console makers, two of which operate consoles in two different console generations, plus cloud-based consoles that use their own technology stacks, namely Nvidia's GeForce Now, Amazon's Luna, and Google's Stadia. There are also two PC platforms, Mac and Windows, that span dozens to hundreds of different hardware builds, and two dominant mobile computing platforms, iOS and Android, that span many more OS versions, GPUs, CPUs, and other chipsets. Every additional platform, device, or build requires code that is unique to a specific set of hardware, or that is written to work across many and without genericizing performance to the lowest common denominator. Creating and supporting all this code is costly, time consuming, and arduous. Another option is to just ignore a large part of the market, which is also expensive.

This challenge, when combined with the ever-growing complexity of virtual words, is why cross-platform game engines such as Unity and Unreal have proliferated. They emerged as a response to fragmentation, and they don't just solve it, they do it at a low cost and to everyone's benefit—even that of the most entrenched platforms.

Imagine a developer decides to build a new game for iOS. Apple's mobile ecosystem has 60% of smartphone share in the United States and 80% share among teenagers, and over two-thirds of mobile gaming revenues globally. In addition, a developer can reach nearly 90% of iOS users by writing to only a dozen iPhone SKUs. The remainder of the global market is split between thousands of different Android devices. Forced to choose between these two platforms, a developer would always pick iOS. But, by using Unity they can easily publish their game across all platforms (including web), thereby increasing their revenue potential by over 50% at little incremental cost.

Apple might prefer more exclusive games, and games fully optimized for their hardware, but it's better for everyone, including iOS users and the App Store, that most mobile developers use Unity. By making more money, developers can build more and better games, thereby driving even more user spending on mobile devices.

The proliferation of cross-platform game engines such as Unity and Unreal should also make it easier to bring together the many fragmented virtual worlds operating today into a unified Metaverse. In fact, this has already been proven. For more than a decade after online console gaming emerged, Sony refused to support cross-play, cross-purchasing, or cross-progression between games played on its PlayStation and other platforms. Sony's policy meant that even if a developer created versions of their game for both PlayStation and Xbox, and two friends bought copies of that same game, they could never play together. Even if a single player bought two copies of the same game (say, one for their PlayStation and another for their laptop), their in-game currencies and many of their rewards would remain siloed in one or the other.

Critics of the policy argued that Sony's stance was a consequence of its dominant market position. The first PlayStation outsold the runner-up console, Nintendo 64, by 200%, and Xbox by over 900%. The PlayStation 2 sold 550% more than the Xbox and Nintendo GameCube combined. The PlayStation 3 barely beat the Xbox 360, largely due to the Xbox's early innovations in online games, and lost to the Nintendo Wii, but by the mid-2010s, the PlayStation 4 had doubled sales of the Xbox One and quadrupled those of the Wii U.

As a result, it seemed that PlayStation saw cross-platform gaming as a threat. If users didn't need a PlayStation to play with other PlayStation users—the majority of console gamers—they'd be less likely to buy a PlayStation in the first place, and PlayStation users might even churn away to competitors. Sony's president of interactive entertainment tacitly admitted as much in 2016, stating "the technical aspect could be the easiest" part of opening up access to its PlayStation Network for cross-play.[2] Yet only two years later, PlayStation enabled cross-play, cross-purchasing, and cross-progression. Three years after that, nearly every game that could support this functionality offered it.

Sony didn't change its mind because of internal preferences, business models, or pressures. Instead, it did so in response to the success of *Fortnite*, which came from a company, Epic Games, that not coincidentally focused on cross-platform gaming.

*Fortnite* had a number of rare attributes when it launched. It was the first mainstream AAA game* that could be played on nearly every major gaming device globally, including two generations of PlayStations and Xboxes, the Nintendo Switch, Mac, PC, iPhone, and Android. The title was also free, which meant players didn't have to buy multiple copies to be able to play on multiple platforms. *Fortnite* was also designed as a social game; it became better as more of your friends used it. And it was built around live services, rather than a fixed narrative or any offline play: the game's content never ended and was updated as often as twice per week. This, plus superb creative execution, helped *Fortnite* become the most popular AAA game globally (excluding China) by the end of 2018. It was generating more revenue per month than any game in history.

Sony's gaming competitors all embraced cross-platform services for *Fortnite*. PC and mobile had never blocked cross-platform functionality; neither Windows nor any mobile platform had ever bought exclusive games. Nintendo supported numerous cross-platform services for *Fortnite* from the start, too—but, unlike Sony, it had no real online networking business and didn't prioritize it. Microsoft, for its part, had long pushed for cross play (likely for the same reason Sony resisted it).

The lack of cross-platform integration meant that PlayStation not only had the worst version of *Fortnite*, but PlayStation owners had many better versions at their fingertips and didn't need to pay a dollar to use them. This fundamentally changed Sony's thinking. Denying such a capability for titles like *Call of Duty* might have had a mod-

---

* "AAA" is an informal classification for video games with large production and marketing budgets and which usually come from the largest video game studios and publishers. It is similar to the "blockbuster" designation in the film industry. Neither term means the title is a financial success.

est impact on the number of copies sold by Activision, but with *Fortnite* Sony was missing out on most of the game's revenue and driving PlayStation players to competing platforms. Sure, PlayStation offered a better technical experience than the iPhone, but most players considered the game's social elements to be more important than its specs. And Epic "accidentally" activated cross-play on PlayStation, allegedly without Sony's permission, on at least three occasions—thereby rallying even more upset users to petition Sony for change, and proving the impediment was policy, not technology.

All of these factors forced Sony to change its policies. This has obviously been for the good of all. Today, a number of hit games can be accessed by nearly all computing devices globally (and thus could be played by anyone, anywhere, anytime), without users needing to repay or fragment their identity, achievements, or player networks. Furthermore, cross-platform play, progression, and purchasing means every console competes on hardware, content, and services. Sony still thrives, too: PlayStation drives over 45% of total *Fortnite* revenues (and the PlayStation 5 has outsold the Xbox Series S and X by a ratio of more than 2:1).[3]

Crucially, Sony's decision to open up its closed platform also provides a view into potential economic solutions to the challenge of interoperability. In order to avoid "revenue leakage," Sony required Epic to "true up" its payments to the PlayStation store. For example, if a *Fortnite* player spent 100 hours playing on PlayStation and 100 hours on Nintendo Switch, but spent only $40 on PlayStation compared to $60 on Nintendo Switch, Epic would have to pay Nintendo a 25% commission on its $60, but then pay PlayStation 25% on its $40 *and* the $10 its share of time would suggest it was owed. In other words, Epic pays twice on that $10. It's not clear whether this policy is still in place—the public only knows it exists because of Epic's lawsuit against Apple. Regardless, the model is an example of how the proliferation of cross-platform gaming aids all market participants.

The success of Discord is another good example. Historically, gaming platforms such as Nintendo, PlayStation, Xbox, and Steam tightly protect their player networks and communication services. This is why someone on Xbox Live cannot "friend" someone on PlayStation Network, nor speak to them directly. Instead, users on other platforms are only available inside cross-platform games, such as *Fortnite*, and via their game-specific IDs. While this approach worked well enough when two players knew which single game they wanted to play before logging on, it didn't work well for unplanned hanging out, or enjoying ad hoc play. The more central gaming was to someone's lifestyle, the less this solution suited them.

Discord emerged to meet this demand, and it has offered gamers numerous benefits. It operates across all major computing platforms— PC, Macs, iPhones, and Androids—meaning every gamer can access a single social graph (and non-gamers can join, too). The service also presents gamers with a rich suite of APIs that can be integrated into other games and even quasi-competitive social services, such as Slack and Twitch, as well as stand-alone games it doesn't distribute or otherwise operate. Discord has been able to build a gamer communication network larger—and far more active—than any single immersive gaming platform.

Importantly, there was no way for the platforms to stop users from using the Discord apps on their phones and using its chat features, in particular. Discord's success drove both Xbox and PlayStation to eventually announce native integrations of Discord into their closed platforms—a move that created a new "interchange" solution for their player networks, communication services, and socializing online.

## Establishing Common
## 3D Formats and Exchanges

The standardization of game engines and communications suites is fairly complex compared to how 3D-objects conventions will emerge.

Witness the current universe of 3D assets. Billions of dollars have been spent on non-standardized virtual objects and environments across film and video games, civil and industrial engineering, healthcare, education, and more. There are no signs that this level of spending will do anything but increase in the near future. Constantly remaking these objects for a new file format or engine is financially impractical and often wasteful; the greatest attribute of a digital "thing" is that it can be endlessly re-used without additional cost.

Interchange solutions are already emerging to tap into the "virtual gold mine" of previously created and fragmented asset libraries. A good example is Nvidia's Omniverse, which launched in 2020 and enables companies to build and collaborate in shared virtual simulations built upon 3D assets and environments from different file formats, engines, and other rendering solutions. An automotive company might be able to bring its Unreal-based cars to an environment designed in Unity and have those cars interact with objects made in Blender. Omniverse doesn't support all possible contributions, nor all metadata and functionality, but because of this, it provides independent developers a clearer reason to standardize. Collaboration, meanwhile, leads to formal and informal conventions. Notably, Omniverse is built on Universal Scene Description (USD), an interchange framework developed by Pixar in 2012 and open sourced in 2016. USD provides a common language for defining, packaging, assembling, and editing 3D, with Nvidia likening it to HTML, but for the Metaverse.[4] In short, Omniverse is driving both an interchange platform and a 3D standard. Helios, the proprietary real-time rendering engine used by the visual effects services company Industrial Light & Magic, is another good example, as it is compatible with only select engines and file formats.

As 3D collaboration grows, standards will naturally emerge. By the early 2010s, for example, globalization had led many of the world's largest corporations to mandate English as their official corporate language—this included Rakuten, Japan's largest e-commerce company; Airbus, an aerospace giant which counts the governments of France and Germany as its two largest shareholders; Nokia, the fourth-largest company in Finland; Samsung, South Korea's largest company, and more. A 2012 poll by Ipsos found that 67% of individuals whose work involved communications with people based in other countries preferred to conduct this work in English. The next closest language was Spanish, at 5%. Crucially, 61% of respondents said they did not use their native language when working with foreign partners, thus the alignment on English was not a reflection of the fact most respondents were primarily English speakers.[5] Globalization has also led to de facto standards in currencies (namely the US dollar and the Euro); units (e.g., the metric system); exchange (the intermodal shipping container); and so on.

Crucially, as Omniverse showed, software does not need everyone to speak the same language. Instead, think of it as comparable to the system within the European Union, which has 24 official languages represented but three (English, French, German) "procedural" languages which are prioritized (also, much of the EU's leadership, parliament, and staff can speak at least two of these languages).

Epic Games, meanwhile, is working to pioneer data standards that allow a single "asset" (really, a right to data) to be repurposed in multiple environments. Not long after acquiring Psyonix, Epic Games announced that the publisher's hit game *Rocket League* would go free-to-play and move over to Epic Online Services. A few months later, Epic announced the first of several "Llama-Rama" events. These limited-time modes enabled *Fortnite* players to complete challenges in *Rocket League* that would unlock exclusive outfits and achievements that could be worn in and across either game. A year later, Epic bought Tonic Games Group, makers of *Fall Guys* and dozens of other games, as part of its investments "in building the Metaverse."[6] It's likely Epic will extend its *Rocket League* experiments across Tonic's titles, as well as those coming from its Epic Games Publishing, which finances and distributes games from independent studios.

With its cross-title assets and achievements model, Epic is likely looking to set a similar precedent as those the company established in cross-platform gaming. Epic clearly believes that there are benefits—

that is, profits—to reducing the friction to accessing different games, making it easier to bring your friends and items across these games, and giving players a reason to try out new games. Players will then spend more time gaming, with more people, across a greater diversity of titles, spending more money along the way. If so, an ever-expanding network of third-party games will want to connect to Epic's virtual identity, communications, and entitlements systems (i.e., parts of Epic Online Services), thereby driving standardization around Epic's various offerings.

Alongside Epic are a series of other socially focused software giants looking to use their reach to establish common standards and frameworks for shared virtual goods. One clear example comes from Facebook, which is adding "interoperable avatars" to its Facebook Connect set of authentication APIs. Facebook Connect is better known to the public as "Log in with Facebook," which allows Facebook users to substitute their Facebook log-in for a website or app's own account system. Most developers would prefer that people create a bespoke account, as it provides the developer with greater information about the user, and means the developer controls this information and account (and not Facebook). However, Facebook Connect is far simpler and faster, and therefore it is the preferred solution for most users. As a result, developers benefit from more registered users (versus anonymous ones). A similar value proposition will exist for Facebook's avatar suite (or perhaps, those of Google or Twitter or Apple). If customized avatars are essential to user expression in 3D space, then few users will want to create a new, detailed avatar for every virtual world they use. The services that accept the avatars a user has already invested in will be able to offer a better experience for said user. Some people even argue that the inability to use a consistent avatar means that no avatar can truly represent the user—just as we wouldn't say that Steve Jobs had a uniform if he could only sometimes wear jeans and a black mock turtleneck, and occasionally needed to wear chambray pants and a gray turtleneck depending on the venue. That's an aesthetic, rather than a uniform intended to reinforce your identity. Regardless, the establishment of cross-title services such as Facebook will serve as another de facto standardization process (in this case, based on Facebook's specifications and advanced by its AR, VR, and IVWP initiatives).

In addition to driving asset interoperability, Epic is also driving the "interoperation" of competing intellectual properties, which is a philosophical, not technical, problem (cross-platform gaming reminds us this is the harder of the two challenges). As virtual platforms like *Fortnite*, *Minecraft*, and *Roblox* grew into culture-driving social spaces, they've become an increasingly necessary part of consumer marketing, brand building, and multimedia franchise experiences. In the past three years, *Fortnite* has produced experiences with the NFL and FIFA, Disney's Marvel Comics, *Star Wars* and *Alien*, Warner Bros.' DC Comics, Lionsgate's *John Wick*, Microsoft's *Halo*, Sony's *God of War* and *Horizon Zero Dawn*, Capcom's *Street Fighter*, Hasbro's G.I. Joe, Nike and Michael Jordan, Travis Scott, and more.

But to participate in these experiences, brand owners must embrace something they almost never allow: unlimited-term licenses (in-game outfits are kept by players forever), overlapping marketing windows (some brand events are mere days apart or overlap entirely), and little to no editorial control. In sum, this means it's now possible to dress as Neymar while wearing a Baby Yoda or Air Jordan backpack, holding Aquaman's trident, and exploring a virtual Stark Industries. And the owners of these franchises *want* this to happen.

If interoperability truly has value, then financial incentives and competitive pressure will eventually solve for it. Developers will eventually figure out how to technically and commercially support Metaverse business models. And they'll use the Metaverse's larger economy to surpass "legacy" game makers.

This is one lesson of the rise of free-to-play game monetization. In this business model, players are charged nothing to download and install a game—or even to play it—but are presented with optional in-game purchases such as an extra level or a cosmetic item. When it was first introduced in the 2000s, and even a decade later, many believed free-to-play would, at best, lead to lower revenues for a given game and at worst cannibalize the industry. Instead, it proved to be the best way to monetize a game and a core driver behind video gaming's cultural ascendance. Yes, it led to many non-paying players, but it substantially increased the total number of players and even gave paying players a reason to spend more. After all, the more people you can show a customized avatar off to, the more you'll pay to do so.

Just as free-to-play led to new products to sell to players, from dances to voice modulators and "battle passes," interoperability will too. Developers might bake degradation into an asset's code—this skin works for 100 hours of use, or 500 games, or three years, during which time it slowly wears out. Alternatively, users might have to pay an additional fee to bring an item from one publisher's title to a competing publisher's (just as many goods have import duties in the "real world") or pay more in the first place for an "interoperable edition." Not all virtual worlds will move to widely interoperable model, of course. Despite the prevalence of free-to-play multiplayer online games today, many titles are still paid, single-player, offline, or all three.

Web3-focused readers might be wondering why I've not yet addressed blockchains, cryptocurrencies, and non-fungible tokens. These three interrelated innovations look likely to play a foundational role in our virtual future and are already operating as a sort of common standard across an ever-expanding series of worlds and experiences. But before examining these technologies, we must first examine the role of hardware and payments in the Metaverse.

# Chapter 9

# HARDWARE

FOR MANY OF US, THE MOST EXCITING ASPECT OF the Metaverse is the development of new devices that we might use to access, render, and operate it. This usually leads to visions of super-powerful, yet lightweight, augmented reality and immersive virtual reality headsets. These devices are not required for the Metaverse, but are often assumed to be the best or most natural way to experience its many virtual worlds. Big-tech executives seem to agree, even though supposed consumer demand for these devices has yet to translate into sales.

Microsoft began developing its HoloLens AR headset and platform in 2010, releasing the first device in 2016 and the second in 2019. After five years on the market, fewer than half a million units have shipped. Still, investment in the division continues and Microsoft CEO Satya Nadella still highlights the device to investors and customers, particularly in the context of the company's Metaverse ambitions.

Although Google Glass, Google's AR device, quickly earned a reputation as one of the most overhyped and failed products in consumer electronics history after it launched in 2013, Google continues to support it. In 2017, the company released an updated model, named the Google Glass Enterprise Edition, with a follow-up coming later in 2019. Since June 2020, Google has spent $1 billion–$2 billion acquiring AR glasses start-ups such as North and Raxium.

Though Google's efforts in VR received less press attention than Google Glass, they've been more significant and arguably more disappointing. Google's first foray came in 2014 and was named Google Cardboard and had the stated goal of inspiring interest in immersive

virtual reality. For developers, Google produced a Cardboard software development kit, which helped developers create VR-specific
apps built in Java, Unity, or Apple's Metal. For users, Google created
a $15 foldout cardboard "Viewer," into which users could place their
iPhones or Android devices in order to experience VR without needing
to buy a new device. A year after Cardboard was announced, Google unveiled Jump, a platform and ecosystem for VR filmmaking, and
Expeditions, a program focused on providing VR-based field trips
for educators. The top-level numbers achieved by Cardboard were
impressive: over 15 million viewers were sold by Google in five years,
while nearly 200 million Cardboard-enabled apps were downloaded,
and over a million students took at least one Expeditions tour within
the first year of release. However, these figures reflected evidence of
consumer intrigue more than inspiration. In November 2019, Google
shut down the Cardboard project and open sourced its SDK. (Expeditions was discontinued in June 2021.)

In 2016, Google launched its second VR platform, Daydream, which
was intended to improve upon Cardboard's foundation. The improvements started with the quality of the Daydream viewer. The $80–$100
headset was made from foam and covered in soft fabric (available in
four colors) and unlike the Cardboard viewer, could be strapped to a
user's head, rather than requiring to user to hold it up in front of them
when in use. The Daydream viewer also came with a dedicated hand-
held remote control, and had an NFC (near-field communication) chip
that could automatically recognize properties of the phone that was
being used and put it into VR mode, instead of requiring users to do so
themselves. While Daydream received positive reviews from the press
and led companies including HBO and Hulu to produce VR-specific
apps, consumers showed little interest in the platform. Google cancelled the project at the same time as Cardboard was terminated.

Despite struggling with AR and VR, Google still appears to view
these experiences as central to its Metaverse strategy. Only a few weeks
after Facebook publicly unveiled its vision of the future in October 2021,
Clay Bavor, Google's head of AR and VR, was made a direct report to

Google/Alphabet CEO Sundar Pichai, and placed in charge of a new group, "Google Labs." It contains all of Google's existing AR, VR, and virtualization projects, its in-house incubator, Area 120, and any other "high-potential long-term projects." According to press reports, Google plans to release a new VR and/or AR headset platform in 2024.

In 2014, Amazon launched its first and only smartphone, the Fire Phone. What differentiated the device from market leaders Android and iOS were the use of four front-facing cameras, which adjusted the interface in response to the user's head movements, and Firefly, a software tool that automatically recognized texts, sounds, and visual objects. The phone turned out to be—and remains—the company's biggest failure, cancelled barely a year after launch. Amazon recognized a $170 million write-down, primarily for unsold inventory. Yet the company soon began work on Echo Frames, a pair of glasses that lacked any sort of visual display, but included integrated audio, Bluetooth (to pair to a smartphone), and the Alexa assistant. The first Echo Frames released in 2019, with an updated model edition released a year later. Neither seems to have sold well.

One of the most outspoken proponents of AR and VR devices is Mark Zuckerberg. In 2014, Facebook acquired Oculus VR for $2.3 billion, more than twice the sum it paid for Instagram two years earlier, even though Oculus had yet to release its device to the public. Not long thereafter, Zuckerberg and his lieutenants mused publicly about the prospect of VR headset PCs as the primary computer for professionals, with wearable AR glasses becoming the primary way that consumers would access the digital world. Eight years later, Facebook announced that the Oculus Quest 2 had sold over 10 million units between October 2020 and December 2021—a figure that beat Microsoft's new Xbox Series S and X console, which released around the same time. However, the device has yet to replace the PC, of course, and Facebook has yet to release an AR device. Still, the bulk of Facebook's $10 billion–$15 billion in annual Metaverse investments are believed to focus on AR and VR devices.

Apple has been, as usual, secretive about its plans for or even belief in AR or VR—but its acquisitions and patent filings are revealing. Over the past three years, Apple has bought start-ups such as Vrvana, which produced an AR headset called Totem; Akonia, which produced lenses for AR products; Emotient, whose machine learning software tracked facial expressions and discerned emotions; RealFace, a facial recognition company; and Faceshift, which remapped a user's facial movements to a 3D avatar. Apple also purchased NextVR, a VR content producer, as well as Spaces, which created location-based VR entertainment and VR-based experiences for videoconferencing software. On average, Apple is granted over 2,000 patents per year (and files for even more). Hundreds of those relate to VR, AR, or body tracking.

Beyond the tech giants, a number of midsize social technology players are investing in proprietary AR/VR hardware, despite having little to no history producing, let alone distributing and serving, consumer electronics. For example, although Snap's first AR glasses, 2017's Spectacles, received more acclaim for their pop-up vending machine sales model than their technical, experiential, or sales success, the company has released three new models over the last five years.

The size of investment in these devices, even in the face of constant rejection from consumers and developers, stems from a belief that history will repeat itself. Every time there is a large-scale transformation in computing and networking, new devices emerge to better suit their capabilities. The companies that first crack these devices, in turn, have the opportunity to alter the balance of power in technology, not just produce a new business line. Thus companies such as Microsoft, Facebook, Snap, and Niantic see the ongoing struggles with AR and VR as proof that they may be able to displace Apple and Google, which operate the most dominant platforms of the mobile era, while Apple and Google understand that they must invest to avoid disruption. There are early signals validating the belief that AR and VR are the next big device technology, too. In March 2021, the US Army announced a deal to buy up to 120,000 customized HoloLens devices from Microsoft over the following decade. This contract was valued at $22 billion— nearly $200,000 per headset (this includes hardware upgrades, repairs, bespoke software, and other Azure cloud computing services).

Another sign that mixed-reality devices are the future is that it's possible to identify numerous technical shortfalls in VR and AR headsets that might be holding up mass adoption. In this sense, some argue that current devices are to the Metaverse what Apple's ill-fated Newton tablet was to the smartphone era. The Newton was released in 1993 and offered much of what we expect from a mobile device—a touchscreen, dedicated mobile operating system and software—but it lacked even more. The device was nearly the size of a keyboard (and weighed even more), could not access a mobile data network, and required the use of a digital pen rather than the user's finger.

With AR and VR, one key constraint is the device display. The first consumer Oculus, released in 2016, had a resolution of 1080 × 1200 pixels per eye, while the Oculus Quest 2, released four years later, had a resolution of 1832 × 1920 per eye (roughly equivalent to 4K). Palmer Luckey, one of Oculus's founders, believes that more than twice the latter resolution is required for VR to overcome pixilation issues and become a mainstream device. The first Oculus peaked at a 90-Hz refresh rate (90 frames per second), while the second offered 72–80 Hz. The most recent edition, 2020's Oculus Quest 2, defaults to 72 Hz, but supports most titles at 90 Hz and provides "experimental support" for 120 Hz on less computationally intensive games. Many experts believe 120 Hz is the minimum threshold for avoiding the risk of disorientation and nausea. According to a report published by Goldman Sachs, 14% of those who've tried an immersive VR headset say they "frequently" experience motion sickness while using the device, 19% respond "sometimes," and another 25% encounter it rarely, but not never.

AR devices have even greater limitations. The average person sees roughly 200°–220° horizontally and 135° vertically, representing a roughly 250° diagonal field of view. The most recent version of Snap's AR glasses, which cost roughly $500, have a 26.3° diagonal field of view—meaning roughly 10% of what you can see can be "augmented"—and runs at 30 frames per second. Microsoft's HoloLens 2, which costs $3,500, has twice the field of view and frame rate, but still leaves 80% of a user's eyesight without augmentation, even though the entirety of their eyes (and much of their head) are covered by the device. HoloLens 2 weighs 566 grams, or 1.25 pounds (the lightest iPhone 13 weighs 174 grams, while the iPhone 13 Pro Max is 240 grams) and supports only two to three hours of active use. Snap's Spectacles 4 weigh 134 grams and can only operate for 30 minutes.

## The Hardest Technology Challenge of Our Time

We might assume that tech companies will inevitably find ways to improve displays, reduce weight, increase battery life, while adding new functionality. After all, TV resolutions seem to increase every year, while supported refresh rates go up, prices go down, and the profile of the device itself narrows. Yet Mark Zuckerberg has said that "the hardest technology challenge of our time may be fitting a supercomputer into the frame of normal-looking glasses."[1] As we saw when examining compute, gaming devices don't just "display" previously created frames, as a TV does—they must render these frames themselves. And as with the challenge of latency, there may be real limitations imposed by the laws of the universe when it comes to what's possible with AR and VR headsets.

Increasing both the number of pixels rendered per frame as well as the number of frames per second requires substantially greater processing power. This processing power also needs to fit inside a device that can be comfortably worn on your head, rather than be stored inside your living room credenza or held in the palm of your hand. And crucially, we need AR and VR processors to do more than just render more pixels.

The Oculus Quest 2 indicates the scale of the obstacle. Like most gaming platforms, Facebook's VR device has a battle royale game, *Population: One*. But this battle royale doesn't support 150 concurrent users, as *Call of Duty Warzone* does, nor 100 like *Fortnite*, or even

*Free Fire*'s 50. Instead, it's limited to 18. The Oculus Quest 2 can't handle much more. Furthermore, the graphics of this game come closer to those of the PlayStation 3, which released in 2006, rather than 2013's PlayStation 4, let alone 2020's PlayStation 5.

We also need AR and VR devices to perform work we don't typically ask from a console or PC. For example, Facebook's Oculus Quest devices include a pair of external cameras that can help alert a user who might otherwise bump into a physical object or a wall. At the same time, these cameras must be able to track user's hands so that they can be re-created inside a given virtual world, or use them as controllers, with given motions or finger movements substituting for the press of a physical button. This might seem a poor substitute for a real controller, but it liberates a VR or AR headset owner from needing to travel with (or walk down the street using) one. Zuckerberg has also spoken to the desire to include cameras on the interior of an AR or VR headset in order to scan and track the user's face and eyes so that the device can steer the user's avatar based on facial and eye movements alone. However, all these additional cameras add more weight and bulk to a headset, while also requiring more computing power, not to mention more battery power. Of course, they add cost, too.

To put this in perspective, we can compare Microsoft's HoloLens 2 to Snap's Spectacles 4. Although the former offers twice the field of view and frame rate of the latter, it is also seven times the price ($3,000–$3,500 versus $500), weighs four times as much, and rather than resemble a pair of futuristic Ray-Bans, looks more like the faceplate and skull of a cyborg. For consumer AR devices to take off, we likely need a device more powerful than the HoloLens, yet smaller than the Spectacles 4. While industrial AR headsets can be larger, they're still constrained by the need to fit under a helmet and the need to minimize neck strain—and must also improve severalfold.

The immense technical challenge of "supercomputer glasses" helps explain how tens of billions of dollars are being spent annually on the problem. But despite this investment, there will be no sudden breakthrough. Instead, there will be a constant process of improvements that reduce the price and size of AR and VR devices, while increasing their computing power and functionality. And even when a key barrier is broken by a given hardware platform or component provider, the rest of the market typically follows within two to three years. What will ultimately differentiate a given platform is the experiences it offers.

We can see this process clearly in the history of the iPhone—the most successful product of the mobile era.

Today, Apple designs many of the chips and sensors inside of its devices, but its first several models were entirely comprised of components created by independent suppliers. The first iPhone's CPU came from Samsung, its GPU from Imagination Technologies, various image sensors were from Micron Technologies, the touchscreen's glass was from Corning, and so on. Apple's innovations were less tangible—how these components were brought together, when, and why.

Most obviously, Apple bet on the touchscreen, skipping a physical keyboard altogether. The move was widely ridiculed at the time, especially by market leaders Microsoft and BlackBerry. Apple also chose to focus on consumers, rather than appealing to large enterprises and small-to-medium businesses, which represented the majority of smartphone sales from the mid-1990s through the late 2000s. Even more radical was the iPhone's price: $500–$600, compared to $250–$350 for competing smartphones, such as the BlackBerry (which were often free to the end user, too, as they were provided by an employer). Apple co-founder and CEO Steve Jobs believed that its $500 device offered superior value—more than a $200 or $300 device ever could—even if the latter was available for free.

Jobs's bets on touchscreens, target market, and price point all proved correct. They were also aided by interface choices that often seemed contradictory, yet perfectly navigated the tension between complexity and simplicity. A good case study is the iPhone's "home button."

Although Jobs showed little interest in physical keyboards, he nevertheless decided to place a large "home button" on the front of the iPhone. The button is now a familiar design element, but it was a novel approach at the time. It also came at a significant cost. The button took up space that might otherwise have been used for a larger screen, longer-lasting battery, or more powerful processor. However, Jobs saw it as an essential part of introducing consumers to both touchscreens and pocket-sized computing. Not unlike closing a flip phone, a user knew that no matter what was happening on their iPhone's touchscreen, pressing the home button would always take them back to the main screen.

In 2011, four years after it released the first iPhone, Apple added a new feature to its operating system: multitasking. Before this point, users could only operate a few pre-determined applications at the same time. It was possible to listen to music through the iPod app while reading the *New York Times* app, but if the user then opened up their Facebook app, the *Times* app would close. If the user wanted to return to a given article in the *Times* app, they'd have to reopen the app, then navigate back to the article and their specific place in it. Doing so would mean exiting the Facebook app, too. Multitasking now allowed users to effectively "pause" an app while switching to another, all of which was managed by the home button. If a user clicked the home button, the app would be paused and they'd return to the home screen. If they double-clicked, the app would still be paused, and a tray of all paused apps would be displayed and could be swiped through.

The first few iPhones could have supported multitasking. After all, other smartphones with similar CPUs supported the feature. However, Apple believed it needed to ease users into the mobile computing era, and this meant focusing not just on what technology was possible, but when users were ready for it. To this end, it wasn't until 2017, with the release of the tenth iPhone, that Apple felt comfortable removing a physical home button and requiring users to "swipe up" from the bottom of the screen instead.

There are no "best practices" within the brand-new device category. In fact, many of the choices we consider obvious today were once controversial—not just the iPhone's touchscreen. For example, some early Android builds and apps used Apple's "pinch-to-zoom" concept, but believed it was backward—if you're bringing your fingers closer together, shouldn't whatever you're looking at come closer, not move farther out? It's almost impossible to imagine this logic today, but that's partly because we've been trained for fifteen years into thinking the opposite is natural. Apple's "slide-to-unlock" feature was considered so novel the company was awarded a patent for it, and ultimately won over $120 million after a US Court of Appeals found that Samsung had violated this patent, among others owned by Apple. Even the app store model was controversial. Smartphone leader BlackBerry didn't launch its app store until 2010, two years after Apple and a year after its famous "There's an app for that" campaign. What's more, BlackBerry's focus on business users (and therefore security) led to policies so strict, such as the need for notarized documents just to gain access to BlackBerry's application development kit, that many developers never even bothered with the platform.

We can already observe echoes of the "smartphone wars" in the VR and AR race. As we've seen, Snap's AR glasses cost less than $500 and target consumers, while Microsoft's run for $3,000 or more and focus on enterprises and professionals. Google believed that rather than sell a multi-hundred- or multi-thousand-dollar VR headset, consumers should instead place the expensive smartphone they already own into a "viewer" that costs less than $100. Amazon's augmented reality glasses don't even have a digital display, and instead emphasize its audio-based Alexa assistant and trendy form factor. Facebook, unlike Microsoft, seems to be focusing on VR before AR, and Zuckerberg and many of his top lieutenants have mused that cloud game streaming may be the only way for a VR user to participate in a richly rendered, high concurrency simulation. Zuckerberg has also said that as a socially focused company, he believes Facebook's AR devices are likely to place a greater emphasis on face- and eye-tracing cameras, sensors, and capabilities as compared to his competitors, who might otherwise focus on minimizing the size of a device, or maximizing its aesthetics. Yet no one knows the exact trade-offs between, say, device profile and functionality, or price and functionality. To capitalize on developer dissatisfaction with Apple's and Google's closed app store models (a topic I'll explore in more detail in the next chapter), Zuckerberg has promised to keep Oculus "open," enabling developers to directly distribute their apps to users and for users to install non-Oculus app stores on their Oculus devices. Though this will surely help attract developers, it produces new risks in user and data privacy—especially as the number of cameras on the device grow.

For AR and VR, it seems clear that the hardware challenges are greater than those for smartphones. And by adapting interfaces from 2D touch to a mostly intangible 3D space, it's likely that inter-face design will also be harder. What will be the "pinch-to-zoom" or "slide-to-unlock" of AR and VR? What, exactly, are users capable of and when?

## Beyond Headsets

Alongside the many investments in immersive headsets are countless other efforts to produce new Metaverse-focused hardware that will complement our primary computing devices, rather than substitute for them, as some imagine AR and VR devices might one day.

Most commonly, gamers imagine wearing smart gloves and even bodysuits that can provide physical (that is, "haptic") feedback to sim-ulate what's happening to their avatar in a virtual world. Many such devices exist today, though they're so costly and functionally limited that they're typically used exclusively for industrial purposes. Specifi-cally, these wearables use a network of motors and electroactive actu-ators that inflate tiny air pockets, thereby applying pressure to their owner or limiting their ability to move.

Haptic vibration technology has advanced considerably since Nin-tendo introduced the Rumble Pak for the Nintendo 64 in 1997. Today's controller triggers, for example, can be programmed with context-specific resistance—a shotgun, sniper rifle, and crossbow will all *feel* different to "pull." The crossbow might even fight back, with the user struggling to hold it down and sensing the vibrations of a virtual bowstring that doesn't really exist.

Another class of haptic interface devices emit ultrasonic sound (that is, mechanical energy waves beyond the audible range for humans) from a grid of microelectromechanical systems (known as MEMS), producing what a user might describe as a "force field" in the air in front of them. The force field produced by these devices, which looks a bit like a short, perforated tin box, is typically less than six or eight inches tall and wide, but its nuance tends to surprise. Test subjects claim to be able to sense everything from a plush teddy bear to a bowling ball and the shape of a sandcastle as it crumbles, aided in part by the fact that fingertips contain more nerve endings than almost any other part of the body. Crucially, MEMS devices can also detect the user's interaction, enabling its sound-based teddy bears to respond to the user's air-based touch, or the castle to crumble if it's touched.

Gloves and bodysuits can also be used to capture a user's motion data, rather than just relay feedback, thereby allowing the wearer's body and gestures to be reproduced in a virtual environment in real time. This information can also be captured using tracking cameras. However, such cameras require unobstructed views, relative proximity to the user, and can struggle if they need to track more than a single user in rich detail. Many users—families, for instance—will want multiple tracking cameras in their "Metaverse rooms" and might add a few smart wearables to their wrists or ankles.

Such bands may seem clumsy (how could a bracelet or anklet substitute for a high-definition camera watching every finger?), but even current technology is impressive. The sensors in an Apple Watch, for example, can distinguish between a user clenching or releasing their fist, between pinching one finger to their thumb and two fingers to their thumb, and can use these movements to interact with the Apple Watch and potentially other devices. In addition, people wearing the watch can use the clenching motion to place a cursor on the Watch face, then use the orientation of their hand to move it. The software involved, Apple's AssistiveTouch, works using fairly standard sensors, including an electric heart monitor, a gyroscope, and an accelerometer.

Other approaches promise even greater capabilities. Facebook's highest price acquisition since Oculus VR in 2014 was CTRL-labs, a neural interface start-up that produces armbands that record electrical activity from skeletal muscles (a technique called electromyography). Although CTRL-labs' devices are worn more than six inches away from the wrist and even farther away from the fingers, CTRL-labs' software enables minute gestures to be reproduced inside virtual worlds—from individual fingers being raised to count, point, or gesture to "come hither," as well as pinches between different sets of fingers. Importantly, CTRL-labs' electromyographic signals can go beyond reproducing human appendages. A famous CTRL-labs demo involves a user—in this case, an employee—mapping their fingers to a crab-like robot and then walking it forward, back, and side-to-side by flexing their fist and moving their fingers.

Facebook is also planning its own line of smart watches. But unlike Apple, Facebook doesn't see the device as secondary to, or dependent upon, a smartphone. Instead, Facebook's watch is intended to have its own wireless data plan and includes two cameras, both of which are detachable and intended to be integrated into third-party items, such as a backpack or a hat. Meanwhile, Google's fifth-largest acquisition ever was the smart wearables company Fitbit, which the company bought for over $2 billion in early 2021.

Wearables will shrink in size and increase in performance, and as the technology improves they will be integrated into our clothes. These developments will help users enhance their interactions with the Metaverse, and enable them to interact with it in more places. Carrying a controller everywhere you go is not practical, and if the primary goal of AR is to make technology disappear into an everyday pair of glasses, then pulling out a thumbstick or a smartphone to use it truly defeats the purpose.

Some believe that the future of computing isn't a pair of AR glasses or a watch, or another kind of wearable, but something smaller. In 2014, only a year after its ill-fated Google Glass launch, Google announced its first Google Contact Lens project, which was intended to help diabetics monitor their glucose levels. Specifically, this "device" was made up of two soft lenses, with a wireless chip, a wireless antenna that's thinner than a strand of human hair, and a glucose sensor placed in between. A pinhole between the underlying lens and the wearer's eyes enabled tear fluid to reach the sensor, which measures blood sugar levels. The wireless antenna drew power from the wearer's smartphone, which was intended to support at least one reading per second. Google also planned to add a small LED light that could warn users of spikes or drops in blood sugar levels in real time.

Google discontinued its diabetes smart lens program four years after it was launched, but the company claimed that the cancellation stemmed from "insufficient consistency in our measurements of the correlation between tear glucose and blood glucose concentration," which had been widely noted by researchers in the medical community. Regardless, patent filings show that the major Western, Eastern, and Southeast Asian technology companies continue to invest in smart lens technology.

Fanciful though such technologies might seem in a world in which internet connections remain unstable and computing scarce, they nevertheless feel within reach compared to so-called brain-to-computer interfaces (BCIs), which have been under development since the 1970s and continue to attract greater investment. Many purported BCI solutions are noninvasive—think of Professor X's helmet in *X-Men*, or perhaps a grid of wired sensors hidden underneath the wearer's hair. Other BCIs are partially invasive or fully invasive, depending on how close the electrodes are placed to brain tissue.

In 2015, Elon Musk founded Neuralink, of which he remains CEO, and announced that the company was working on a "sewing machine–like" device which could implant sensors that are four-to-six micrometers thick (roughly 0.000039 inches, or one-tenth the width of human hair) into the human brain. In April 2021, the company released a video in which a monkey played the game *Pong* using a wireless Neuralink implant. Only three months later, Facebook announced that it was no longer investing in its own BCI program. In prior years, the company had funded several projects inside and outside the company, including a test at the University of California, San Francisco, which involved wearing a helmet which shot light particles through the skull, then measured blood oxygenation levels in groups of brain cells. A blog post on the subject explained that "while measuring oxygenation may never allow us to decode imagined sentences, being able to recognize even a handful of imagined commands, like 'home,' 'select,' and 'delete,' would provide entirely new ways of interacting with today's VR systems—and tomorrow's AR glasses."[2] Another Facebook BCI test involved a physical mesh of electrodes placed on top of the user's skull, which enabled the subject to write at a speed of roughly 15 words per minute purely through thought (the average person types 39 words per minute, two and a half times faster). Facebook reported that "while we still believe in the long-term potential of head-mounted optical [brain-computer interface] technologies, we've decided to focus our immediate efforts on a different neural interface approach that has a nearer-term path to market for AR/VR"[3] and that "a head-mounted optical silent speech device is still a very long way out. Possibly longer than we would have foreseen."[4] The "differential neural interface approach" that Facebook referred to is likely CTRL-labs, but part of the problem with BCI's "path to market" is ethics, not tech. How many people want a device that can read their thoughts—and not just the thoughts that are related to the task at hand? Especially if that device is permanent?

# The Hardware Around Us

In addition to the devices we hold, wear, and maybe even implant as part of our transition to the Metaverse, there are the devices that will proliferate throughout the world around us.

In 2021, Google unveiled Project Starline, a physical booth designed to make video conversations feel like you're in the same room with the other participant. Unlike a traditional monitor or videoconferencing station, Starline's booths are powered by a dozen depth sensors and cameras (together producing seven video streams from four viewpoints and three depth maps), as well as a fabric-based, multilayered light-field display, and four spatial audio speakers. These features allow participants to be captured and then rendered using volumetric data, rather than flattened 2D video. During internal tests, Google found that in comparison to typical video calls, Starline users focused 15% more on those they're speaking to (based on eye-tracking data), displayed significantly greater non-verbal forms of communication (e.g., ~40% more hand gestures, ~25% more head nods, and ~50% more eyebrow movements), and had 30% better memory recall when asked to remember details from their conversation or meeting.[5] The magic, as always, is in the software, but it depends on the extensive hardware to be realized.

Storied camera manufacturer Leica now sells $20,000 photogrammetric cameras that have up to 360,000 "laser scan set points per second," and that are designed to capture entire malls, buildings, and homes with greater clarity and detail than the average person would ever see if they were physically on-site. Epic Games' Quixel, meanwhile, uses proprietary cameras to generate environmental "MegaScans" comprising tens of billions of pixel-precise triangles. The satellite imaging company Planet Labs, mentioned in Chapter 7, performs scans of nearly the entire earth daily across eight spectra bands, enabling not just daily high-resolution imagery, but details including heat, biomass, and haze. To produce this imagery, it operates the second-largest fleet of satellites of any company in the world,* with over 150, many of which weigh less than 5 kilograms and are smaller than 10 × 10 × 30 centimeters. Every photo from these satel-

---

* As points of comparison, China has fewer than 500 satellites, while Russia has fewer than 200. However, they are typically far larger and more capable than those of Planet Labs.

lites covers 20–25 square kilometers and is made up of 47 megapixels, with each pixel representing 3 × 3 meters. Roughly 1.5 GB of data is sent from each of these satellites per second and from an average distance of 1,000 kilometers. Will Marshall, the CEO and co-founder of Planet Labs, believes that the cost per performance of these satellites has improved by 1,000 times since 2011.[6] Such scanning devices make it easier and cheaper for companies to produce high-quality "mirror worlds" or "digital twins" of physical spaces—and to use scans of the real world to produce higher-quality and less-expensive fantasy worlds.

Also important are real-time tracking cameras. Consider Amazon's cashier-less, cashless, and auto-pay grocery stores, Amazon Go. These stores deploy scores of cameras that track every customer by way of facial scanning as well movement tracking and gait analysis. A customer can pick up and put down whatever they like, then simply walk out of the store, having paid for only what they took with them. In the future, this sort of tracking system will be used to reproduce these users, in real time, as digital twins. Technologies such as Google's Starline might simultaneously allow workers to be "present" in the store (potentially from an offshore "Metaverse call center"), jumping across different screens to help the customer.

Hyper-detailed projection cameras will also play a part, enabling virtual objects, worlds, and avatars to be transplanted into the real world and in realistic detail. Key to these projections are various sensors that enable the cameras to scan and understand the non-flat, non-perpendicular landscapes they will project against, and alter their projection accordingly so that it appears undistorted to the viewer.

Technologists have long imagined an internet-of-things future where sensors and wireless chips are as ubiquitous as electrical outlets, albeit more diverse, thereby enabling us to light up any number of experiences wherever we go. Imagine a construction site with drones overhead, each filled with cameras, sensors, and wireless chips, and with workers below them wearing AR headsets or glasses. This setup would allow a site operator to know exactly what is happening, where, and at all times, including the total volume of sand in a given mound, the number of trips needed to move it by machine, who is closest to a problem area and best able to address it, when, and with what impact.

Of course, not all of these experiences require the Metaverse, or even virtual simulation. However, humans find 3D environments and data presentation far more intuitive—consider the difference between seeing a digital tablet containing the status of a job site and instead seeing that information overlaid on top of the site and its objects. Notably, Google's second-largest acquisition ever (the largest if we exclude Motorola, which Google divested after three years) is that of Nest Labs, which develops and operates smart sensor devices, for $3.2 billion in 2014. Eight months after the acquisition, Google spent another $555 million to acquire Dropcam, a smart-camera maker, which was then folded into Nest Labs.

## Long Live the Smartphone?

It's fun to imagine all the brilliant new devices that will soon enough allow us to jack into the Metaverse. But, at least through the 2020s, it's likely that most devices for the Metaverse era will be those we already use.

Most experts, including Unity Technologies CEO John Riccitiello, estimate that by 2030, there will be fewer than 250 million active VR and AR headsets in use.[7] Of course, betting on such long-term prognostications is dangerous. The first iPhone was released in 2007, eight years after the first BlackBerry smartphone and at a point when smartphone penetration was less than 5% in the US. Eight years later, the iPhone had sold more than 800 million units and driven US penetration to nearly 80%. Few in 2007 believed that by 2020, two-thirds of everyone on earth would have a smartphone.

Still, AR and VR devices face not just significant technical, financial, and experiential obstacles, but also the need to squeeze in. Behind the rapid growth of smartphones was a simple pair of facts: the personal computer was one of the most significant inventions in human history, but more than 30 years after it was invented, less than one in six people worldwide owned one. And those lucky few who did? Well, their computers were large and immovable. AR and VR devices will not be a person's first computing device, nor even their first portable one. They are fighting to be a person's third, or even fourth—and for a long time yet will probably be one of their least powerful, too.

AR and VR may come to replace most of the devices we use today. That time is unlikely to be soon. Even if the combined number of VR and AR headsets (two very different device types) in use by 2030 tops one billion, four times the aforementioned forecast, they would still reach fewer than one in six smartphone users. And that's okay. There are, in 2022, hundreds of millions of people spending hours each day inside real-time rendered virtual worlds through smartphones and tablets—and these devices are rapidly improving.

In earlier sections, I reviewed the ongoing improvements in smart-phone CPUs and GPU power. These are probably the most important Metaverse-related advancements to these devices, but they're far from the only ones worth highlighting. Since 2017, new iPhone models have included infrared sensors that track and recognize 30,000 points on the user's face. While this functionality is most commonly used for Face ID, Apple's face-based authentication system, it also enables app developers to reproduce a user's face in real time as an avatar or with virtual augmentations. Examples include Apple's own Animoji, Snap's AR lenses, and Epic Games' Unreal-based Live Link Face app. In the years to come, many virtual worlds operators will use this capability to let players map their facial expressions to their in-world avatars—live and without requiring another dollar in hardware.

Apple has also led the deployment of lidar scanners into smartphones and tablets.* As a result, even most engineering professionals no longer see the need to buy $20,000–$30,000 lidar-specific cameras, and close

---

* Lidar determines the distance and shape of objects by measuring the time it takes for a reflected laser (i.e., light beam) to return to a receiver, similar to how radar scanners use radio waves.

to half of American smartphone users can now create and share virtual-izations of their homes, offices, yards, and everything inside them. This innovation has transformed companies such as Matterport (discussed in Chapter 7), which now produces thousands of times the number of scans per year, and with far greater diversity.

The iPhone's high-resolution, three-lens cameras also enable users to create high-fidelity virtual objects and models using photographs, with the asset stored in the Universal Scene Description interchange framework. These objects can then be transplanted into other virtual environments—thereby reducing the cost and increasing the fidelity of synthetic goods—or overlaid into real environments for the purpose of art, design, and other AR experiences.

Oculus VR, meanwhile, uses the high-resolution, multi-angle iPhone camera to produce mixed reality experiences. For example, an Oculus user playing *Beat Sabre** can place their iPhone several meters behind them, so they can see themselves inside a VR environment from within their VR headset, and all from a third-person perspective.

Many new smartphones also feature new ultra-wideband (UWB) chips that emit up to 1 billion radar pulses per second and receivers that process the return information. Smartphones can thereby create extensive radar maps of the user's home and office, and know exactly where the user is within these maps (or others such as Google's street or building maps), and relative to other users and devices. Unlike GPS, UWB offers precision down to a few centimeters. The front door to your home can automatically unlock when you approach from the out-side, but know that when you're tidying the shoe rack indoors, the door should not unlock. Using a live radar map, you'll be able to nav-igate much of your home without ever needing to remove your VR headset—your device will alert you a potential collision, or render the potential obstacle inside your headset so that you can move around it.

That all of this is possible through standard consumer-grade hard-

---

* *Beat Sabre* is like *Guitar Hero*, though notes are hit not by pressing a button on a physical keyboard, but by hitting a virtual button with a virtual lightsaber.

ware is astonishing. The growing role of this functionality in our daily lives explains why the average sales price of Apple's iPhone has increased from roughly $450 in 2007 to over $750 in 2021. Put another way, consumers have not asked Apple to use the cost-side of Moore's Law to offer the capabilities of the first few iPhones but at lower price. Nor have they asked Apple to use the performance-side of Moore's Law to improve the prior year's iPhone while maintaining its price. Instead, consumers want more—more of, well, nearly everything the iPhone might do.

Some believe that the future role of the smartphone includes operating as a user's "edge computer" or "edge server," providing connectivity and compute to the world around us. Versions of this model already exist. For example, most of the Apple Watches sold today lack a cellular network chip and instead connect to their owner's iPhone via Bluetooth. The approach has limitations: the Apple Watch cannot make a phone call when taken too far away from its tethered iPhone, nor play music to a wearer's AirPods, download new apps, retrieve messages not already stored on the watch, and so on. But in exchange, the device is much cheaper, lighter, and less battery intensive—all because the user's iPhone, a far more powerful device, and one with greater cost per performance, is doing most of the work.

Similarly, the iPhone will send complex Siri queries to Apple's servers for processing, while many users choose to store the bulk of their photos in the cloud instead of buying iPhones with larger hard drives, which can cost between $100 and $500 more. Earlier, I mentioned that many believe VR headsets must at least double the screen resolution offered by top-of-the line devices in the market today and reach 33%–50% higher frame rates (that is, produce more than two and a half times the number of pixels per second) if they're to gain mainstream adoption. Not only that, but this must happen alongside cost reductions, shrinking the device's profile, and minimizing heat. While the technology doesn't exist yet in a single device, by connecting to a sufficiently powered PC via Oculus Link, the Oculus Quest 2 can reliably hike its frame rate while also increasing rendering capabilities. In

January 2022, Sony announced its PlayStation VR2 platform, which boasted 2000 × 2040 pixels per eye (roughly 10% more than the Oculus Quest 2) and a 90–120-Hz refresh rate (compared to the OQ2's 72–120), with 110° field of view (versus 90°), plus eye tracking (not available). However, the PSVR2 requires users to own and physically connect to Sony's PlayStation 5 console, which costs more than the cheapest Oculus Quest 2, and does not come with the PSVR2 headset.

Given the shortage, importance, and cost of compute, it makes sense to focus on a single device's capabilities, rather than invest in scores of others, especially when those devices have greater physical, thermal, and cost limitations. A computer worn on your wrist or face simply cannot contend with one in your pocket. This logic applies to more than just compute. If Facebook wants us to wear a CTRL-labs band on each limb, why load each one with their own cellular and WiFi network chips, if cheaper, less energy intensive, and smaller Bluetooth chips are available, and they can send that data to a smartphone to manage? Personal data may be the most important consideration. We probably don't want our data being collected, stored, or sent to a wide network of devices. Instead, most of us would prefer that this data is sent from these devices to the one we trust most (and one that's stored on our person), and for that device to manage which other devices can have access to other parts of our online histories, information, and entitlements.

## Hardware as the Gateway

The many devices required and expected to support the Metaverse can be grouped into three categories. First, the "primary computing devices," which for most consumers are smartphones, but may be AR or immersive VR at some point in the future. Second, the "secondary" or "supporting computing devices," such as a PC or PlayStation, and likely AR and VR headsets. These devices may or may not rely on a primary device, or be complemented by them, but they will be used less frequently than a main device and for more specific purposes. Finally, we have the tertiary devices, such as a smartwatch or tracking camera, which enrich or extend a Metaverse experience, but will rarely operate it directly.

Each category and sub-category of devices will increase Metaverse engagement time and total revenues—and offer manufacturers an opportunity to generate a new business line. However, the massive investment in these devices—many of which are years from mainstream viability—has broader motivations.

The Metaverse is a mostly intangible experience: a persistent network of virtual worlds, data, and supporting systems. However, physical devices are the gateway to accessing and creating these experiences. Without them there is no forest to be known, heard, smelled, touched, or seen. This fact provides the device manufacturers and operators with significant soft and hard power. The manufacturers and operators will determine what GPUs and CPUs are used, the wireless chipsets and standards deployed, the sensors included, and so on. Critical though these intermediary technologies are to a given experience, they rarely interface directly with a developer or end user. Instead, they're accessed through an operating system, which manages how, when, and why the capabilities are used by a developer, which sorts of experiences they're allowed to offer a user, and whether and to what extent a commission must be paid to the maker of said device.

In other words, hardware is not just about what the Metaverse might offer and when, but a fight to influence how it works and, ideally, to take a cut of as much of its economic activity as possible. The more important the device—and the more devices that connect to it—the greater the control afforded to the company which makes it. To understand what this means in practice, we need to dive deep into payments.

# Chapter 10

# PAYMENT RAILS

THE METAVERSE IS ENVISIONED AS A PARALLEL plane for human leisure, labor, and existence more broadly. So it should come as no surprise that the extent to which the Metaverse succeeds will depend, in part, on whether it has a thriving economy. Yet we are not accustomed to thinking in these terms; while science fiction has predicted the Metaverse, one usually finds only glancing references to a virtual world's internal economy in such stories. A virtual economy might sound like a strange, daunting, even confusing prospect, but it shouldn't be. With a few important exceptions, the Metaverse economy will follow the patterns of real-world ones. Most experts agree on many of the attributes that produce a thriving real-world economy: rigorous competition, a large number of profitable businesses, trust in its "rules" and "fairness," consistent consumer rights, consistent consumer spending, and a constant cycle of disruption and displacement, among others.

We can see these attributes at play in the world's largest economy. The United States was not built by any one government or corporation, but instead by millions of different businesses. Even in today's era of megacorporations and technology giants, the country's more than 30 million small-to-medium businesses employ over half the workforce and are responsible for half of GDP (both figures exclude military and defense spending). Amazon's hundreds of billions in sales are almost exclusively of things other companies make. Apple's iPhone is one of the most significant products in human history and each year, Apple designs an ever-growing portion of its richly integrated components.

However, most of its components still come from competitors—and many of them are constantly warring with Apple over prices, while enabling the company's competitors. In addition, consumers buy (and frequently upgrade to new versions of) this incredible device in order to access content, apps, and data made in large part by companies other than Apple.

Apple is a prime example of the dynamism of the American economy. Although the company was an early leader in the PC era of the 1970s and 1980s, it struggled throughout the 1990s as Microsoft's ecosystem grew and internet services proliferated. But through the iPod in 2001, iTunes in 2003, iPhone in 2007, and App Store in 2008, Apple became the most valuable company in the world. It isn't difficult to imagine another outcome: one in which Microsoft, whose operating system powered 95% of the computers used to manage an iPod or run iTunes, was able to stymie its would-be competitor as a means of buttressing its Windows Mobile and Zune offerings. Alternatively, we could imagine a version of earth in which internet providers such as AOL, AT&T, or Comcast had been able to use their power over data transmission to control what content could flow through their systems, how, and with which royalties.

The American economy is backed by an elaborate legal system which covers everything made or invested in, what's sold and bought, who is hired and what tasks they perform, and also what's owed. While this system is imperfect, costly to use, and often slow, its existence instills in all market participants faith that their agreements will be honored and that there is some middle ground between "free market competition" and "fairness" which benefits all parties. The success of Apple, as well as other internet giants that were born during the PC era, such as Google and Facebook, is inextricably linked to the famous court case *United States v. Microsoft Corporation*, which found the company to have unlawfully monopolized its operating system through the control of APIs, forced bundling of software, restrictive licensing, and other technical restrictions. Another example is the "first sale doctrine," which allows someone who buys a copy of a copyrighted work from the copyright holder to dispose of that copy however they wish. This is why Blockbuster was able to purchase a $25 VHS tape and then endlessly rent it to its customers without needing to pay royalties to the Hollywood studio that made it, and why you have the right to sell your copy of a book or rip up and restitch a shirt with a copyrighted design.

In this book so far, I've examined many of the innovations, conven-

| Largest Public Corporations by Market Capitalization (Excluding State-Owned Corporations) *in Trillions of Dollars* | | | | | |
|---|---|---|---|---|---|
| **March 31, 2002** | | | **January 1, 2022** | | |
| 1 | General Electric | $0.372 | 1 | Apple | $2.913 |
| 2 | Microsoft | $0.326 | 2 | Microsoft | $2.525 |
| 3 | Exxon Mobil | $0.300 | 3 | Alphabet (Google) | $1.922 |
| 4 | Walmart | $0.273 | 4 | Amazon | $1.691 |
| 5 | Citigroup | $0.255 | 5 | Tesla | $1.061 |
| 6 | Pfizer | $0.249 | 6 | Meta (Facebook) | $0.936 |
| 7 | Intel | $0.204 | 7 | Nvidia | $0.733 |
| 8 | BP | $0.201 | 8 | Berkshire Hathaway | $0.669 |
| 9 | Johnson & Johnson | $0.198 | 9 | TSMC | $0.623 |
| 10 | Royal Dutch Shell | $0.190 | 10 | Tencent | $0.560 |

*Sources: "Global 500," Internet Archive Wayback Machine, https://web.archive.org/web/20080828204144/http://specials.ft.com/spdocs/FT3BNS7BW0D.pdf; "Largest Companies by Market Cap," https://companiesmarketcap.com/.*

tions, and devices required to achieve a flourishing and fully realized Metaverse. But I have not yet addressed one of the most important: "payment rails."

Because most payment rails largely predate the digital age, we tend not to think of them as "technology." In truth, they are the embodiment of digital ecosystems—complex series of systems and standards, deployed across a wide network and in support of trillions of dollars in economic activity, and in a primarily automated fashion. They're typically difficult to build and even harder to displace, and also quite profitable. Visa, MasterCard, and Alipay are each among the 20 most valuable public companies in the world, with most of their peers the likes of Google, Apple, Facebook, Amazon, and Microsoft, as well as large financial conglomerates such as JPMorgan Chase and Bank of America which hold trillions in deposits and manage the transfer of trillions more in financial instruments each day.

Unsurprisingly, there is already a fight to become the dominant "payment rail" in the Metaverse. What's more, this fight is arguably the central battleground for the Metaverse, and potentially its greatest impediment, too. To unpack Metaverse payment rails, I'll first provide an overview of the major payment rails of the modern era, before explaining the role of payments in today's gaming industry and how that role has informed the payment rails of the mobile computing era. Then, I'll discuss how mobile payment rails are being used to control emergent technologies and stifling competition; before touching on why so many Metaverse-focused founders, investors, and analysts see blockchains and cryptocurrencies as the first "digitally native" payment rail and the solution to the problems plaguing the current virtual economy.

## The Major Payment Rails Today

Over the past century, the number of distinct payment rails has grown as a result of new communication technologies, the increase in the number of transactions made per person per day, and the fact that most purchases are not in physical cash. From 2010 to 2021, cash's share of US transactions fell from over 40% to roughly 20%.

The most common payment rails in the United States are Fedwire (formerly known as the Federal Reserve Wire Network), CHIPS (Clearing House Interbank Payment System), ACH (Automated Clearing House), credit cards, PayPal, and peer-to-peer payment services like Venmo. These rails have different requirements, merits, and demerits, having to do with the fees that are charged, network size, speed, reliability, and flexibility. We'll come back to this later on when I discuss blockchains and cryptocurrencies, so it's important to remember these categories and related details.

Let's start with the classic payment rail: wires. In the mid-1910s, the US Federal Reserve Banks began to move funds electronically, eventually establishing a proprietary telecommunications system that spanned each of the 12 Reserve Banks, the Federal Reserve Board, and the US Treasury. Early systems were telegraphic and used Morse code, but by the 1970s, Fedwire began to move to telex, then computer operations, followed by proprietary digital networks. Wires can only be used between (and thus through) banks, so both the sender and receiver must have a bank account. For similar reasons, a wire can only be sent on non-holiday weekdays and during business hours. While a sender can set up recurring wires (say, sending $5,000 every Tuesday), there is no such thing as a "wire request." As such, wires cannot be used to automatically pay recurring bills or other invoices. Once money is sent via wires, that sending cannot be reversed. Even if this were possible, other limitations discourage frequent use of wires. For example, there are often significant fees charged to both the sender ($25–$45) and recipient ($15), in addition to other fees for non-USD wires, failed wires, confirmations (which are not always possible), and more. The banks themselves are charged as little as $0.35 and $0.9 per transaction by Fedwire. The size of these fees, which are mostly fixed, makes small dollar-value wires impractical. But for larger sums (individuals can wire up to $100,000), wires are the least expensive option.

In the 1970s, the major US banks also established a competitor to (and customer of) Fedwire, named CHIPS, in part to reduce their transfer costs. In particular, this meant departing from Fedwire's "real-time settlement," in which a sender's wire would be instantly received by and usable to the recipient. In contrast, each bank holds outgoing CHIPS wires until the end of the day, at which point they are grouped together based on the recipient bank, and then netted against all of the incoming CHIPS wires from that same bank. In simple terms, CHIPS means that rather than Bank A sending Bank B millions of wires per day, and Bank B sending Bank A millions of wires per day, they wait until the end of the day and make a single transaction. With this system, neither the sender of a wire nor its recipient has access to the wire's funds (and for up to 23 hours, 59 minutes, and 59 seconds). Only the bank does, and the bank collects interest on much of it over the course of the day. Naturally, banks typically default to CHIPS for their wires. Due to time zones, money-laundering protections, and other governmental restrictions, international wires typically take two to three days.

As anyone who has sent a wire knows, they are typically the most complex and time-consuming way to send money, as extensive information is needed from the recipient. The irreversibility of the transaction, combined with the lack of (or delayed) confirmation, also means that mistakes are even more time-consuming to correct. Yet wires are still typically considered the most secure way to send money, as CHIPS is limited to only 47 member banks and involves no intermediary, while Fedwire's only intermediary is the United States Federal Reserve. In 2021, $992 trillion in value was sent via US Fedwire across 205 million transactions (an average of roughly $5 million), while CHIPS cleared over $700 trillion across an estimated 250 million transactions (an average of $3 million).

An ACH is an electronic network for processing payments. The first ACH emerged in the United Kingdom in the late 1960s. Like wires, ACH payments can only be made during business hours, and they require the sender and the recipient to each have a bank account.

These bank accounts must usually be part of the ACH network, and as such ACH payments face geographic limitations, in most cases. A Canadian bank account can typically make an ACH payment to one in the United States, but making an ACH payment to Vietnam, Russia, or Brazil is unlikely to be possible, or at least requires various intermediaries that increase costs. The fees associated with ACH payments are seen as their primary differentiator. Most banks allow customers to make ACH-based transfers for free, or at most, for $5. Businesses can make ACH payments to vendors or employees for less than 1% per transaction. Unlike a wire, an ACH payment is also reversible and allows payment requests from would-be recipients. These capabilities, mixed with its low cost, is why this rail is typically used to make payments to vendors and employees, and set on "auto-pay" for electrical, telephone, insurance, and other bills. An estimated $70 trillion was processed by the US ACH in 2021, spanning more than 20 billion transactions (an average of roughly $2,500 per transaction).[1]

The major downside of ACH is that it is slow: transactions take anywhere from one to three days. This is because ACH payments don't "clear" until the end of the day (some banks perform a few batches per day), at which point a bank aggregates everything it must send to another bank (that is, all ACHs) and sends it in one sum via Fedwire, CHIPS, or similar solution. The resulting lag produces several challenges beyond just one to two and a half days during which neither sender nor receiver has the funds. For example, with ACH there's no confirmation of a successful transaction—you're only notified if there is an error. And that error takes several days to correct: the recipient bank doesn't notice the failure until the second day, its report isn't processed until the end of the second day, and the original sender receives the notice the following day (at which point the three-day process starts again).

Makeshift credit card systems have existed since the late 19th century, though what we now consider a "credit card" did not emerge until the 1950s. Today, we "swipe" or "tap" a physical card (or enter our credit card information into a box online), after which a credit card machine or remote server captures the account information and digitally sends it to the merchant's bank, which then submits it to the customer's credit card provider, which either grants or denies the transaction. The process takes one to three days, though the consumer of course doesn't notice this, and typically costs merchants 1.5% to 3.5% of the transaction. This fee is much higher than that for an ACH payment, but credit cards enable a transaction to be placed in seconds and without exchanging detailed and personal bank account information. The purchaser does not even need to have a bank account.

While credit cards are often free to the user, late payments and interest can quickly result in paying more than 20% per year on top of the relevant transactions (it's likely you pay your credit card bill via ACH). Credit card operators generate a third of their revenue through other services sold to merchants and credit card owners, such as insurance, or by selling data generated in their network. Like ACH, but unlike wires, credit card payments can be reversed, though this process can take days, is often contested, and is only available for a few hours or days after a transaction (a dispute can be filed far later). Like wires, credit cards work across almost all markets globally. And, unlike both wires and ACH, credit card payments are supported by almost all merchants, and transactions can be placed at any time and every day. As anyone with a credit card knows, they are typically the least secure way to make a payment, and often suffer the most fraud. An estimated $6 trillion was spent via credit card in the US in 2021, with an average of $90 across more than 50 billion transactions.

Finally, there are the digital payment networks (also known as peer-to-peer networks) such as PayPal and Venmo. Although users do not need bank accounts to open a PayPal or Venmo account, these accounts must be funded, with the money coming from an ACH payment (a bank account), credit card payment, or transfer from another user. Once funded, these platforms then serve as a centralized bank used by all accounts; transfers between users are effectively just reassignments of money held by the platform itself. As a result, payments are instantaneous and can be made irrespective of the day or time. When money is sent between friends and family, these platforms typically do not charge a fee. However, payments made to businesses typically involve fees of 2%–4%. And if a user wants to move their on-platform money to their bank account, they must typically pay 1% (up to $10) for it to arrive the same day, or otherwise wait two to three days (during which the platform collects interest). Finally, these networks are usually geographically limited (Venmo is US only, for example) and do not support peer-to-peer payments outside their networks (that is, a PayPal user cannot send funds to a Venmo wallet, meaning they must instead route it through several intermediary accounts or rails). In 2021, an estimated $2 trillion was processed globally by PayPal, Venmo, and Square's Cash App, with an average of roughly $65 per transaction across more than 30 billion transactions.

In sum, the various US payment rails tend to vary in terms of security, fees, and speed. No payment rail is perfect, but more important than their technical attributes is that they compete with one another, including within each category. There are multiple wire rails, multiple credit card networks, multiple digital payment processors and platforms. Each of these competes based on their advantages and drawbacks, and even within a single category, there are various fee strucutres. The credit card operator American Express, for example, charges far more than Visa, but it offers consumers more lucrative points and perks, and merchants a higher-income clientele. If a user decides they don't want a credit card, or a merchant declines to take Amex, each has multiple alternatives available. And again, they can also make some transfers for free if they're willing to lend their money to a given digital payment network for two to three days.

# The 30% Standard

We might assume that the virtual world would have "better" payment rails than the "real world." After all, its economy primarily involves goods that only exist virtually, and that are bought via purely digital (and thus low marginal cost) transactions, and are, for the most part, $5–$100 each. The virtual economy is also large. In 2021, consumers spent more than $50 billion on digital-only video games (in contrast to physical discs), and nearly $100 billion more on in-game goods, outfits, and extra lives. As a point of comparison, $40 billion was spent at the theatrical film box office in 2019, the last year before the COVID-19 pandemic, and $30 billion on recorded music. What's more, the "GDP" of the virtual world is growing rapidly—it has quintupled on an inflation-adjusted basis since 2005. In theory, these facts should mean creativity, innovation, and competition in payments. In practice, the reverse has been true; the payment rails of today's virtual economy are more expensive, cumbersome, slow to change, and less competitive than those in the real world. Why? Because what we consider to be a virtual payment rail, such as PlayStation's wallet, Apple's Apple Pay, or in-app payment services, are really a stack of different "real world" rails and forced bundle of many other services.

In 1983, the arcade manufacturer Namco approached Nintendo about publishing versions of its titles, such as *Pac-Man*, on its Nintendo Entertainment System (NES). At the time, the NES was not intended to be a platform. Instead, it played only titles made by Nintendo. Eventually, Namco agreed to pay Nintendo a 10% licensing fee on all of its titles that appeared for NES (Nintendo would have right of approval over every individual title), plus another 20% for Nintendo to manufacture Namco's game cartridges. This 30% fee ultimately became an industry standard, replicated by the likes of Atari, Sega, and PlayStation.[1]

Forty years later, few people play *Pac-Man* anymore, and costly cartridges have transformed into low-cost digital discs manufactured by game-makers and even lower-cost bandwidth for digital downloads (where the costs are mostly borne by consumers via internet fees and console hard drives). However, the 30% standard has endured and expanded to all in-game purchases, such as an extra life, digital backpack, premium pass, subscription, update, and more (this fee also covers the two to three percentage points charged by an underlying payment rail, such as PayPal or Visa).

Console platforms have several rationales for the fee beyond simply making money. The most important is how they enable game developers themselves to make money. For example, Sony and Microsoft typically sell their respective PlayStation and Xbox consoles for less than they cost to manufacture, which makes it cheaper for consumers to access powerful GPUs and CPUs, as well as other related hardware and components that are needed to play a game. And this per-unit loss is before these platforms allocate the research and development investments to design their consoles, marketing costs to convince users to buy them, and exclusive content (that is, Microsoft and Sony's internal game development studios) that encourages users to buy them upon release, rather than years later. Given that new consoles usually enable new or better capabilities, speedier adoption should benefit developers and players alike.

The platforms also develop and maintain a number of proprietary tools and APIs that developers need in order to make their games run on a given console. The platforms also operate online multiplayer networks and services such as Xbox Live, Nintendo Switch Online, and PlayStation Network. These investments help game-makers, but they lead the platform to attempt to recoup and then profit from their expenditures—hence the 30% fee.

Gaming platforms may have a rationale for a 30% fee, but that does not mean the fee is set by the market, nor that it is fully earned. Consumers are forced to buy these consoles at below cost; there's no option to buy a more expensive unit that has 30% lower prices on software. And while consoles must attract developers, they don't compete against one another for these developers. Most game-makers release their titles on as many platforms as possible in order to reach as many players as possible. As such, none of the major consoles stands to benefit from offering developers better terms. A 15% reduction by Xbox would mean a game publisher would make 21% more from every copy they sold on the Xbox, but if they chose not to release their title on PlayStation or Nintendo Switch as a result, they'd miss out on as much as 80% of total sales. This might net Microsoft some additional customers, but not 400% more, the figure required to make the publisher whole. If Microsoft's maneuver was matched by PlayStation or Nintendo, all three platforms would lose half of their software revenue and for little benefit.

The most pointed critique of the 30% cut focuses on the console's proprietary tools, APIs, and services. In many cases, they add cost to the developer, rather than aid them. In other cases, they produce limited value. And in some cases, they serve only to lock in customers and developers alike, and to the detriment of both groups. This reality can be seen clearly in three areas: API collections, multiplayer services, and entitlements.

For a game to run on a specific device, it needs to know how to communicate with that device's many components, such as its GPU or microphone. To support this communication, console, smartphone, and PC operating systems produce "software development kits" (or SDKs) that include, among other things, "collections of APIs." In theory, a developer could write their own "driver" to talk to these components, or use free and open-source alternatives. OpenGL is another collection of APIs used to speak to as many GPUs as possible from the same codebase. But on consoles and Apple's iPhone, a developer can only use those made by the platform's operator. Epic Games' *Fortnite* must use Microsoft's DirectX collection of APIs to speak to the Xbox's GPU. The PlayStation version of *Fortnite* must use PlayStation's GNMX, while Apple's iOS requires Metal, Nintendo Switch requires Nvidia's NVM, and so on.

Each platform argues that their proprietary APIs are best suited to their proprietary operating systems and/or hardware, and therefore developers can make better software using them, which leads to happier users. This is generally true, though the majority of virtual worlds operating today—and especially the most popular ones—are made to run on as many platforms as possible. As such, they are not richly optimized to any platform. Furthermore, many games don't need every ounce of computing power. The variations in API collections and lack of open alternatives are partly why developers use cross-platform game engines such as Unity and Unreal, as they are designed to speak to every API collection. To this end, some developers might prefer to give up a little bit of performance optimization in order to instead optimize for their budget by using OpenGL, rather than pay or share a portion of revenues with Unity or Epic Games.

The challenge of multiplayer is a bit different. In the mid-2000s, Microsoft's Xbox Live managed almost all of the "work" for an online game: communications, matchmaking, servers, and so on. Though this work was hard and expensive, it also substantially increased gamer engagement and happiness, which was good for developers. Yet 20 years later, almost all this cost is now borne and managed by a game's maker. The transition reflects the growing importance of online services, and the shift to support cross-play. Most developers now want to manage their own "live ops," such as content updates, competitions, in-game analytics, and user accounts, and it doesn't make sense for Xbox to manage the live services for a game that's integrated into PlayStation, Nintendo Switch, and more. But game developers are still obligated to pay the full 30% to gaming platforms and work through their online account systems. Furthermore, if the Xbox Live network goes offline due to technical difficulties, as an example, players cannot access *Call of Duty: Modern Warfare*'s online play. And of course, players themselves are already paying a monthly subscription fee to Microsoft for Xbox Live, and no part of that fee goes to the developers whose games justify its existence and who pay the most in server bills.

Critics argue that the real goals of platform services are to create extra distance between developers and players, lock both groups into hardware-based platforms, and justify the platform's 30% fee. So when players buy a digital copy of *FIFA 2017* from the PlayStation Store, that copy is forever tied to PlayStation. In other words, PlayStation has already netted its $20 from a $60 purchase, but if the player wants to play the game on Xbox, they'll have to spend another $60 even if the developer would be willing to give it to the player for free. The more a user pays a console manufacturer like Sony—thus repaying their console loss—the more expensive it becomes to ever leave.

Platforms take a similar approach when it comes to game-related content. If a player beats *Bioshock* on PlayStation, and later switches to Xbox, they not only need to repurchase the game, they'd need to beat it a second time to replay the final level. In addition, if PlayStation awarded *Bioshock* players any trophies (say, for beating the game faster than 99% of other players), PlayStation would keep these awards for eternity. As I discussed in Chapter 8, Sony was able to use their control over online play to prevent cross-platform gaming for more than a decade. This helped neither developers nor players—it obviously harmed both—but it (theoretically) helped Sony retain PlayStation customers by making it harder to acquire Xbox customers.

The payment rails of console gaming are not discrete, as they are in the real world. Players and developers alike are prohibited from directly using credit cards, ACH, wires, or digital payment networks, and the billing solution offered by a platform is bundled with many other things—entitlements, save data, multiplayer, APIs, and more. It doesn't matter what the market rate is, or what a developer or user needs. There is no discount if a publisher's game is offline only, or if they don't need the online multiplayer services of a given platform. It also doesn't matter if a publisher's game was bought at GameStop, rather than digitally at PlayStation's Store—even though the publisher had to give GameStop a cut of the transaction, too. The fee is the fee. The best illustration of this reality is a platform that lacks any hardware at all, yet has proven more dominant than Nintendo, Sony, or Microsoft.

## The Rise of Steam

In 2003, the game-maker Valve launched the PC-only application Steam, effectively the iTunes of games. At the time, most PC hard drives could store just a few games at a time—a problem that was only getting worse as the size of the average game file grew faster than affordable storage space. Finding and then downloading these games, uninstalling them to free up space for others, reinstalling the old game later on when the user wanted to return to it, and shifting them to a new PC were all laborious. A user had to manage multiple credentials, numerous credit card receipts, web addresses, and so on. Furthermore, many online multiplayer games, such as Valve's own *Counter-Strike*, were moving to a "games-as-a-service" model in which the game would be updated or patched on a frequent basis. This allowed games to be "refreshed" with new features, weapons, modes, and cosmetics, but it also meant that players had to constantly update their games, causing no small amount of frustration. Imagine coming home after a long day of work to play *Counter-Strike*, only to discover you had to wait an hour for an update to download and be installed.

Steam solved these problems by creating a "game launcher" that indexed and centrally managed game installer files, but also took care of a user's rights to these games and automatically downloaded and updated the games a player had installed on their PC. In exchange, Steam would keep 30% from the sale of every game through its system—just like the console game platforms.

Over time, Valve added more services to Steam, collectively called Steamworks. For example, Valve used the Steam account system to create an early "social network" of friends and teammates that any game could access. Players no longer had to search for and re-add their friends (or rebuild their teams) every time they bought a new game. Steamworks' Matchmaking, meanwhile, enabled developers to use Steam's player networks to create balanced and fair online multi-player experiences. Steam Voice allowed players to speak in real time. These services were provided at no additional cost to developers, and unlike console platforms, Steam did not charge players themselves to access online networks or services, either. Later, Valve made Steamworks available to games not sold on Steam, such as a physical copy of *Call of Duty* that was bought from GameStop or Amazon, thereby building a bigger and more richly integrated network of online gaming services. Steamworks was theoretically free to developers, but it also forced each game to use Steam's payment service for all subsequent in-game transactions. As such, developers paid for Steamworks by netting Steam 30% of ongoing revenues.

Steam is seen as one of the most important innovations in PC gaming history, and a critical reason the segment remains as large as console gaming, even with its greater complexity of use and higher cost of entry (a decent gaming PC still costs more than $1,000, while meeting the specifications of newer consoles requires $2,000 or more). But nearly 20 years later, its technical innovations in game distribution, rights management, and online services have largely been commoditized. In some cases, users and publishers skip them altogether. Many PC gamers, for example, now use Discord for audio chat, rather than Steam's voice chat. The rise of cross-platform gaming also means that most in-game trophies and play records are awarded and managed by a game-maker, rather than by Steam.

Yet no one has managed to compete with or disrupt Valve's platform, even though PCs, unlike consoles, are open ecosystems. A player can download as many software stores as they like and even buy a game directly from the publisher. The publisher can also withhold that title from Steam and still reach its customers. But Steam's power and centrality endure.

In 2011 gaming giant Electronics Arts launched its own store, EA Origin, which would exclusively sell PC versions of its titles (thus cutting distribution fees from 30% to 3% or less). Eight years later, EA announced it would return to Steam. Activision Blizzard, the studio behind hits such as *Warcraft* and *Call of Duty,* has spent 20 years trying to leave Steam, but except for free-to-play titles such as *Call of Duty: Warzone,* most of its titles continue to sell through the platform. And Amazon, the largest e-commerce platform in the world and owner of Twitch, the largest video game livestreaming service outside China, has struggled to gain any meaningful share of PC gaming—even after it began adding free games and in-game items to its popular Prime subscription. None of the above even prompted a modest fee cut or policy change by Valve.

Steam's ongoing success is partly due to its outstanding service and rich feature set. It is also protected by its forcible bundling of distribution, payments, online services, entitlements, and other policies—just like consoles.

One example is that any game purchased through the Steam store or running through Steamworks will forever require Steam to be played. Even decades after Steam rendered its services to a player and developer, the platform will continue to get a cut of ongoing revenues. The only way around this was for the publisher to pull their game from Steam altogether—which would mean requiring users to repurchase the title through another channel. Because Steam does not allow players to export their achievements earned on the platform, they'd lose any awards issued through Steamworks if they did leave Steam.

According to some reports, Steam also uses "most favored nations" (MFN) clauses to ensure that even if a competing store offered lower distribution fees, a publisher would not be able to exploit this to undercut Steam's consumer-facing prices. Consider a $60 game sold by Steam, which takes $18 (30%) of the $60 and nets $42 to the publisher. If a competitor offered 10% fees, a publisher could still sell that game for $60, thereby netting $54 ($8 more). However, users won't jump from a store they love (and one that is used by all of their friends and contains decades of game purchases and awards) for nothing. A competing store would have to disrupt Steam by splitting the fee reduction with developers and consumers. The game might be sold for $50, which nets $45 to the publisher ($3 more) and saves the consumer $10 (this price cut might result in more total purchases, too). Unfortunately, Steam's MFNs made this impossible. If a publisher cuts their price on a competitor's store, they would have to do the same on Steam. Alternatively, they could leave the store—but a publisher would doubtlessly lose more in customers than they might hope to make up in margin. Crucially, this MFN agreement applied even to a publisher's own store, rather than just third-party aggregators like Steam.

The most notable effort to compete with Steam came from Epic Games, which launched the Epic Games Store in 2018 with the explicit purpose of reducing distribution fees in the PC gaming industry. To attract developers as well as users, Epic sought to offer all of the benefits of Steam, but with fewer limitations and better prices.

Games sold through EGS would not require a player to keep using

EGS as long as they wanted to play the game. Players actually owned a copy of a game, rather than the right to a copy of that game inside EGS; game-makers could therefore leave the store at any time without abandoning their customers. Players owned their in-game data, too. If they ever wanted to leave the platform for a publisher's own store, or any other, they could take their trophies and player networks with them. EGS offered 12% store fees (which dropped to 7% if the developer was already using Unreal, thereby ensuring that even if a developer used Epic's engine and store, they'd pay no more than 12% combined, even if multiple distinct products were bought, used, or licensed).

Epic also used its hit game *Fortnite*, which was generating more revenue per year than any other game in history, to bring players to the store. With an update, PC copies of the game were transformed into the Epic Games Store itself, with *Fortnite* a launchable title within it. Epic also spent hundreds of millions giving away free copies of hit games such as *Grand Theft Auto V* and *Civilization V*, and hundreds of millions more on exclusive windows to a series of not-yet-released PC titles. Due to Steam's MFNs, it could not, however, offer lower prices on non-exclusive titles.

On December 3, 2018—only three days before Epic launched its store—Steam announced that it would cut its commission to 25% after a publisher's title exceeded $10 million in gross sales, and to 20% after $50 million. This was an early victory for Epic, though the company noted Valve's concession most strongly benefited the largest game developers—that is, the few global giants most likely to start their own stores or pull their games from Steam. It did not apply to the many thousands of independent developers struggling to stay afloat, let alone turn a big profit. Valve also declined to open up Steamworks. Nevertheless, the move shifted hundreds of millions in annual profits from Steam to developers.

By January 2020, Epic had spent huge amounts of money yet inspired no further concessions from Steam (nor the console platforms). However, Epic's CEO, Tim Sweeney, expressed his view that competing stores would need to cut their rates, tweeting that EGS was a "coin toss": "Heads, other stores don't respond, so Epic Games Store wins [by stealing market share] and all developers win. Tails, competitors match us, we lose our revenue sharing advantage, and maybe other stores win, but all developers still win."[3] Sweeney's gambit may ultimately prove right, but as of February 2022, Valve's policies had yet to budge a second time. EGS, meanwhile, was accumulating enormous losses and showing limited evidence of sustainable success with players. Epic's public disclosures showed the platform's revenues grew from $680 million in 2019[4] to $700 million in 2020[5] and $840 million in 2021.[6] However, 64% of this spend was on *Fortnite*, with the title also driving 70% of platform revenue growth over the three-year period. With nearly 200 million unique users in 2021, some 60 million of which were active in December, EGS does seem popular (Steam has an estimated 120–150 million monthly users). But as the platform's revenues suggest, many of these players are likely using EGS just to play *Fortnite*, which can only be accessed through EGS on PCs. It's also likely that many non-*Fortnite* players use EGS solely for its free games. In 2021 alone, Epic released 89 titles for free, worth a combined $2,120 at retail (or roughly $24 each). Over 765 million copies were redeemed that year, representing a notional value of $18 billion, compared to $17.5 billion the prior year and $4 billion in 2019.* While these giveaways did draw players, they did not lead to much user spending (they probably harmed it). The average user spent between $2 and $6 on non-*Fortnite* content throughout the entirety of 2021 (and received $90 to $300 in free games). Leaked documents from Epic Games suggested that EGS lost $181 million in 2019, $273 million in 2020, and would lose between $150 and $330 million in 2021, with breakeven occurring in 2027 at the earliest.[7]

One could argue that because PCs are an open platform, no store can have a monopoly—and notably, the dominant online game distributor is independent from both Microsoft and Apple, which run the Windows and Mac operating systems and offer their own stores. At the same time, it's telling that there is only one major profitable store,

---

* Epic pays a deeply discounted wholesale to publishers, with 2021's payments estimated at roughly $500 million.

and its biggest suppliers struggle to exist outside of it. Few should consider this a healthy outcome, especially at a 30% fee or even 20% fee. This is because, as always, payments are a bundle that spans not just the processing of a transaction, but a user's online existence, their storage locker, their friendships, and their memories, as well as a developer's obligation to their oldest customers.

## From Pac-Man to iPod

You might be wondering what *Pac-Man* cartridges, Steam MFNs, and *Call of Duty* copies have to do with the Metaverse. Well, the gaming industry isn't just informing the creative design principles and building the underlying technologies of the "next-generation internet." It also serves as the Metaverse's economic precedent.

In 2001, Steve Jobs introduced digital distribution to most of the world through the iTunes music store. For his business model, he chose to emulate the 30% commission commanded by Nintendo and the rest of the gaming industry (though unlike consoles, the iPod itself had gross margins above 50%, not below 0%). Seven years later, this 30% was transposed to the iPhone's app store, with Google quickly following suit for its Android operating system.

Jobs also decided, at this point, to adopt the closed software model used by the console platforms, but that had not been previously used by its Mac laptops and computers, or its iPod.* On iOS, all software and content would need to be downloaded from Apple's App Store, and as with PlayStation, Xbox, Nintendo, and Steam, only Apple had a say over what software could be distributed and how users would be billed.

---

* While most iPod users bought their music from iTunes, they could also import tracks bought from other services, uploaded from CDs, or even pirated from services such as Napster. More technically savvy users could even download these tracks to an iPod without using iTunes.

Google took a more permissive approach with Android, which technically allowed users to install apps without using the Google Play Store—and without third-party app stores. But this required users to navigate deep into their account settings, and grant permission for individual applications (for example, Chrome, Facebook, or the mobile Epic Games Store) to install "unknown apps," while warning users this made their "phone and personal data more vulnerable to attack" and forcing them to agree "that you are responsible for any damage to your phone or loss of data that may result from their use." Although Google did not take responsibility for any damage or loss of data resulting from the use of apps distributed by its Google Play store, the additional steps and warnings meant that while most PC users downloaded software directly from its maker, such as Microsoft Office from Microsoft.com or Spotify from Spotify.com, almost no one did on Android.

It took more than a decade for the problems associated with the proprietary model employed by Apple, and in a different way, by Google, to surface on the global stage. In June 2020, the European Union sued Apple after Spotify and Rakuten, two streaming media companies, alleged Apple used its fees to benefit its own software services (such as Apple Music) and stifle competitors. Two months later, Epic Games sued both Apple and Google, alleging their 30% fees and controls were unlawful and anti-competitive. A week before the suit, Sweeney had tweeted that "Apple has outlawed the metaverse."

The delay had several causes. One was the unequal impact of Apple's store policies, which primarily charged "new economy" businesses and waived fees on old economy ones. Apple established three broad categories of apps when it came to in-app purchases. The first category was transactions made for a physical product, such as buying Dove soap from Amazon or loading a Starbucks gift card. Here, Apple took no commission and even allowed these apps to directly use third-party payment rails, such as PayPal or Visa, to complete a transaction. The second category was so-called reader apps, which included services that bundle non-transactional content (for example, an all-you-can-eat Netflix, *New York Times*, Spotify subscription), or that allow a user to access content they previously purchased, such as a movie previously bought from Amazon's website that the user now wants to stream on Amazon's Prime Video iOS app. The third category was interactive apps in which users can affect the content (in a game, or cloud drive, for example) or make individual transactions for digital content (such as a specific movie rental or purchase on the Prime Video app). These apps had no choice but to offer in-app billing.

While these interactive apps could offer online browser-based payment alternatives, like reader apps, players could still not be told about these options inside the app itself. As such, these alternatives were rarely used—if known. Imagine the last time you used an app which supported in-app payments from Apple—did you ever wonder whether the app's developer offered better prices online? And if they did, how much cheaper would they need to be for you to bother signing up for their account and entering your payment information, rather than just click "Buy" in the App Store? 10%? 15%? How big would the purchase need to be (saving 20% on a $0.99 extra life doesn't seem worth it)? Maybe 20% worked for most purchases, but then a developer was "saving" only 7%, as they then needed to cover the fees charged by PayPal or Visa. If, instead, a game could require a customer to go elsewhere, like Netflix or Spotify, they might be able to save 20% or even 27%.

Various emails and documents from Epic's court case against Apple revealed that the App Store's multi-category payment models resulted primarily from where Apple believed it could exert leverage. But leverage also correlated with where Apple believed it could create value. Mobile commerce, of course, has been critical to the growth of the global economy for some time, but most of it was a reallocation from physical retail. For many people, the iPad form factor made reading the *New York Times* more compelling on the tablet than in print, but Apple didn't enable the journalism industry. Mobile gaming was different. When the App Store launched, the gaming industry generated just over $50 billion per year—$1.5 billion of which was in mobile. By 2021, mobile was more than half the $180 billion industry and represented 70% of growth since 2008.

The economics of the App Store exemplify this dynamic. In 2020, an estimated $700 billion was spent using iOS apps. However, less than 10% of this was billed by Apple. Of this 10%, nearly 70% was for games. Put differently, seven in every 100 dollars spent inside iPhone and iPad apps were for games, but 70 of every 100 dollars grossed by the App Store were from the category. Given that these devices are not gaming-focused, are rarely bought for this purpose, and that Apple offers almost none of the online services of a gaming platform, this figure often comes as a surprise. The judge overseeing Epic Games' lawsuit against Apple famously told Apple CEO Tim Cook: "You don't charge Wells Fargo, right? Or Bank of America? But you're charging the gamers to subsidize Wells Fargo."[8]

Because the App Store's revenue came primarily from a tiny, but fast-growing, segment of the world economy, it also took time for the App Store to become a large business worth scrutinizing. Ironically, even Apple seemed to doubt it would become one. Two months after its launch, Jobs reviewed the nascent business with the *Wall Street Journal*. In its report, the paper stated that "Apple wasn't likely to derive much in the way of a direct profit from the business. . . . Jobs is betting applications will sell more iPhones and wireless-enabled iPod touch devices, enhancing the appeal of the products in the same way music sold through Apple's iTunes has made iPods more desirable." To this end, Jobs told the *Journal* that Apple's 30% fees were intended to cover credit card fees and other operating expenses for the store. He also said that the App Store "is going to crest a half a billion, soon . . . Who knows, maybe it will be a $1 billion marketplace at some point in time." The App Store passed this $1 billion mark in its second year, with Apple noting that it now operated "a bit over break-even."[9]

By 2020, the App Store had become one of the best businesses on earth. With revenues of $73 billion and an estimated 70% margin, it would've been large enough to be a member of the *Fortune* 15 if it was spun off from its parent company (which is the largest company in the world by market capitalization, as well as the most profitable in dollar terms). And this is despite the fact that the App Store billed less than 10% of transactions flowing through its system, which themselves made up less than 1% of the global economy. Were iOS an "open platform," these profits would likely have been competed away, at least in part. Visa and Square would offer smaller in-app fees, while competing app stores would emerge that offered services comparable to Apple's but at lower prices. But this isn't possible because Apple controls all of the software on its device, and like gaming consoles, keeps it closed and bundled. And its only major competitor, Google, is just as happy with the state of play.

These issues aren't exclusive to the Metaverse, of course, but their consequences for it will be profound, for the same reason Judge Gonzalez Rogers narrowed in on Apple's gaming policies: the entire world is becoming game-like. That means it's being forced into the 30% models of the major platforms.

Take Netflix, as an example. In December 2018, the streaming service chose to remove in-app billing from its iOS app. As a "reader app," this was the company's right, and its financial planning team had decided that while asking users to sign up on Netflix.com and manually enter their credit card would cost them some sign-ups versus Apple's one-click in-app alternative, this missed revenue was less than the 30% that it would have to send Apple.* But in November 2021, Netflix added mobile games to its subscription plan, which turned the company into an "interactive app" and forced the company to return to Apple's own payment service (or stop offering an iOS app altogether).

---

* In 2016, Apple offered subscription apps a drop to 15% commission when a customer reached the second continuous year of a subscription (that is, the 13th month). While this seems significant, as most subscriptions hope to retain subscribers forever, meaning the 30% would apply to only a small portion of customers, the reverse is true. Netflix, for example, has roughly 3.5% monthly churn. This means the average customer lasts for 28 months, which would mean an average 21.5%. Put another way, only 62% of subscribers ever reach a second year. Also, most subscriptions services aren't Netflix. Industry average churn in subscription online video is roughly 6%, or an average of 17 months of service per subscriber, less than 48 of every 100 sign-ups reaching a second year.

But why, exactly, does Apple's 30% "outlaw" the Metaverse, to return to Sweeney's pre-lawsuit remark? There are three core reasons. First, it stifles investment in the Metaverse and adversely affects its business models. Second, it cramps the very companies that are pioneering the Metaverse today, namely integrated virtual world platforms. Third, Apple's desire to protect these revenues effectively prohibits many of the most Metaverse-focused technologies from further development.

## High Costs and Diverted Profits

In the "real world," payment processing costs as little as 0% (cash), typically maxes out at 2.5% (standard credit card purchases), and sometimes reaches 5% (in the case of low-dollar-value transactions with high minimum fees). These figures are low because of robust competition between payment rails (wire versus ACH, for example) and within them (Visa versus MasterCard and American Express).

But in the "Metaverse," everything costs 30%. True, Apple and Android do provide more than just payment processing—they also operate their app stores, hardware, operating systems, suite of live services, and so on. But all of these capabilities are forcibly bundled and consequently not exposed to direct competition. Many payment rails are also bundles. For example, American Express provides consumers with access to credit, as well as its payment networks, perks, and insurance, while merchants gain access to lucrative clientele, fraud service, and more. Yet they are also available unbundled and compete based on the specifics of these bundles. In smartphones and tablets, there is no such competition. Everything is bundled together, in only two flavors: Android and iOS. And neither system has an incentive to cut fees.

This doesn't necessarily mean the bundle is overpriced or problematic. But they certainly appear to be. The average annual interest rate on unsecured credit card loans is 14%–18%, while most states have usury prohibitions that cap rates at 25%. Even the most expensive malls in the world don't charge rents that work out to 30% of a business's revenue, nor do the tax rates in the highest-taxed nations' highest-taxed states' highest-taxed cities come close to 30%. If they did, every consumer, worker, and business would leave and every taxing body would suffer as a result. But in the digital economy, there are only two "countries" and both are happy with their "GDP."

Furthermore, average small-to-medium business profit margins in the US are between 10% and 15%. In other words, Apple and Google collect more in profit from the creation of a new digital business or digital sale than those who invested (and took the risk) to make it. It's hard to argue that this is a healthy outcome for any economy. Considered another way, cutting the commissions of these platforms from 30% to 15% would more than double the profits of independent developers—with much of that money then reinvested into their products. Many if not most would agree that this is probably better than funneling more money to two of the richest companies on earth.

The current dominance of Apple and Google also leads to undesirable economic incentives. Nike, which is already pioneering virtual athletics apparel in the Metaverse, serves as a good example. If Nike sells physical shoes through its Nike iOS app, Apple collects a 0% fee. Later, if Nike decides to give the purchasers of its real-world shoes the rights to virtual copies ("buy Air Jordan's in-store, get a pair in *Fortnite*," for example), Apple will still not take a fee. If the owner then "wears" these virtual shoes in the real world, as might be rendered through an iPhone or forthcoming Apple AR headset, Apple is still owed nothing. The same applies if Nike's physical shoes have Bluetooth or NFC chips that speak to Apple's iOS devices. But if Nike wants to sell stand-alone virtual shoes to a user, or virtual running tracks, or virtual running lessons, Apple is owed 30%. In theory, Apple would be owed a cut if it determined the primary source of value in a combined virtual + physical set of shoes was virtual too. The upshot is a lot of chaos for a series of outcomes in which the function of Apple's device, components, and capabilities is largely the same.

Here is another hypothetical, this time focused on Activision, a virtual-first company, unlike Nike. If a *Call of Duty: Mobile* user buys a $2 pair of virtual sneakers for her character, Apple collects $0.60. But if Activision asks the user to instead watch $2 worth of advertisements in exchange for a free pair of virtual sneakers, Apple collects $0. In short, the consequences of Apple's policies will shape how the Metaverse is monetized, and who leads that process. For Nike, the 18% differential between Apple's 30% fee and the 12% argued by Epic is nice, but not necessary. And if Nike wants, it can skip it altogether by leveraging its existing, physical business. Most start-ups do need the extra margin, however, and can't rely on a pre-Metaverse business line.

These problems are only going to grow in the years to come. Today, a high school tutor can sell video-based lessons directly to customers via web browser, and if they choose to offer an iOS app, they can opt against in-app payments. This is because video-focused apps are "reader apps." But if this tutor wants to add interactive experiences, such as a physics class that involves the construction of a simulated Rube Goldberg machine, or an instructional course on automotive engine repairs with rich 3D immersion, they are obligated to support in-app payments because they're now an "interactive app." Apple or Android receive a cut specifically because this tutor chose to invest in a harder, and more expensive, lesson.

Apple would argue that the added benefit of immersion would justify their cut, but the math here is tricky. A $100 non-interactive textbook sold outside the app store would need to charge $143 to make up for Apple's fee. The teacher would need an even higher price to recoup their added investments and risk—and for every additional dollar they charged, Apple would take 30 cents. At $200, Apple receives $60 for the new lesson, while the teacher's take-home has increased by only $40 and the students are out an extra $100. It's difficult to read this as a positive societal outcome—especially given that the student's educational experience is unlikely to have doubled in quality, no matter the significance of 3D-specific enhancements.

# Constrained Virtual World
# Platform Margins

The problems of 30% payment rails are particularly acute in virtual world platforms.

*Roblox* is full of happy users and talented creators. However, few of these creators are making money. Although Roblox Corporation had nearly $2 billion in revenues in 2021, only 81 developers (i.e., companies) netted over $1 million that year, and only seven crossed $10 million. This is bad for everyone, really, given that more developer revenue would mean more developer investment and better products for users, which in turn drives more user spending.

Unfortunately, it's difficult for developers to increase revenues because Roblox pays them only 25% of every dollar spent on their games, assets, or items. While this makes Apple's 70%–85% payout rates seem generous, the reverse is true.

Imagine a hypothetical involving $100 in Roblox iOS revenue. Based on fiscal 2021 performance, $30 goes to Apple off the top, $24 is consumed by Roblox's core infrastructure and safety costs, and another $16 is taken up by overhead. This leaves a total of $30 in pretax gross margin dollars for Roblox to reinvest in its platform. Reinvestment spans three categories: research and development (which makes the platform better for users and developers), user acquisition (which increases network effects, value for the individual player, and revenues for developers), and developer payments (which leads to the creation of better games on Roblox). These categories receive $28, $5, and $28 (this exceeds Roblox's target 25% due to incentives, minimum guarantees, and other commitments to developers), or $60 combined. As a result, Roblox currently operates at a roughly –30% margin on iOS. (Roblox's blended margin is a bit better at –26%. This is because iOS and Android represent 75%–80% of total revenues by platform, with most of the remaining coming from platforms such as Windows, which do not take a fee).

To summarize, Roblox has enriched the digital world and turned hundreds of thousands of people into new digital creators. But for every $100 of value it realizes on a mobile devices, it loses $30, developers collect $25 in net revenue (that is, before all of their development costs), and Apple collects roughly $30 in pure profit even though the company puts nothing at risk. The only way for Roblox to increase developer revenues today is to deepen its losses or halt its R&D, which would in turn harm both Roblox and its developers over the long term.

Roblox's margins should improve over time as neither overhead nor sales and marketing expenses are likely to grow as fast as revenues. However, these two categories will unlock only a few percentage points—not enough to cover its sizable losses or to marginally increase developer revenue shares. R&D should offer some scale-related margin improvements, too, but fast-growing companies shouldn't be achieving profitability through R&D operating leverage. Roblox's largest cost category, infrastructure and safety, is unlikely to decrease as it is mostly driven by usage (which in turn drives revenue) and if anything, the company's R&D is likely to enable experiences that cost *more* to operate per hour (for example, virtual worlds with high concurrency or that involve more cloud data streaming). The second-largest (and only remaining) cost category is store fees, which Roblox has no control over.

To Apple, Roblox's margin constraints (and the consequences of those constraints on Roblox developer revenues) are a feature, not a bug, of the App Store system. Apple does not want a Metaverse comprised of integrated virtual world platforms, but of many disparate virtual worlds that are interconnected through Apple's App Store and the use of Apple's standards and services. By depriving these IVWPs of cash flow, while offering developers much more of it, Apple can nudge the Metaverse to this outcome.

Let's return to my earlier example of a tutor looking to produce interactive classes. The tutor needs to increase the price of their lesson by 43% or more just to break even due to Apple's 30% cut. But if they shift to Roblox, their price would need to increase by *400%* to offset the 75.5% collected by Roblox and Apple combined. While

Roblox is much easier to use than Unity or Unreal, takes on many additional costs for the tutor (for example, server fees), and aids in customer acquisition, the enormity of this price gap will drive most developers to release standalone apps using Unity and Unreal, or to bundle together in an education-specific IVWP. In either outcome, Apple becomes the primary distributor of virtual software, with the App Store providing discovery and billing services.

## Stopping Disruptive Technologies

The policies of Apple and Google limit the growth potential not only of virtual world platforms, but also the internet at large. For many, the World Wide Web is the best "proto-Metaverse." Though it lacks several components of my definition, it is a massively scaled and interoperable network of websites, all running on common standards and available on nearly every device, running any operating system, and through any web browser. Many in the Metaverse community thus believe that the web and web browser should be the focal point of all Metaverse development. Several open standards are already being shepherded, including OpenXR and WebXR for rendering, WebAssembly for executable programs, Tivoli Cloud for persistent virtual spaces, WebGPU, which aspires to provide "modern 3D graphics and computation capabilities" inside a browser, and more.

Apple has frequently argued that its platform isn't closed because it provides access to the "open web"—that is, websites and web apps. As such, developers need not produce apps to reach its iOS users, especially if they disagree with Apple's fees or policies. Furthermore, the company argues, most developers choose to make apps despite this alternative, which shows that Apple's bundled services are outcompeting the entirety of the web, rather than being anti-competitive.

Apple's argument isn't convincing. Recall the story I highlighted at the beginning of this book, about what Mark Zuckerberg once called Facebook's "biggest mistake." For four years, the company's iOS app was really just a "thin client" that ran HTML. That is, its app had very little code and was, for the most part, just loading various Facebook webpages. Within one month of switching over to an app that was "rebuilt from the ground up" on native code, users were reading double the number of Facebook News Feed stories.

When an app is written natively for a given device, programming is specifically configured for that device's processors, components, and so forth. As a result, the app has more efficient, optimized, and consistent performance. Web pages and web apps cannot directly access native drivers. Instead, they must speak to a device's components through a "translator" of sorts and with more generic (and often bulkier) code. This leads to the opposite outcome of native applications: inefficiency, sub-optimization, and less reliable performance (such as crashes).

But as much as consumers prefer native apps for everything from Facebook to the *New York Times* and Netflix, they're essential to rich real-time rendered 2D and 3D environments. These experiences are computationally intensive—far more so than rendering a photo, loading a text article, or playing back a video file. Web-based experiences largely preclude rich gameplay such as that of *Roblox*, *Fortnite*, and *Legend of Zelda*. This happens to be one of the reasons Apple was able to place such strict in-app billing rules on the gaming categories.

What's more, the web must be accessed through a web browser, which is an application. And Apple uses its control over its App Store to prevent competing browsers on its iOS devices. This may be surprising if you regularly use Chrome on your iPhone or iPad. However, these are really just the "iOS system version of [Apple's Safari] WebKit wrapped around Google's own browser UI," according to the Apple expert John Gruber, and the iOS Chrome app [cannot] "use the Chrome rendering or JavaScript engines." What we think of as Chrome on iOS is simply a variant of Apple's own Safari browser, but one that logs into Google's account system.[*10]

---

* Apple typically forces third-party browsers to use older, and thus slower and less capable, versions of WebKit than iOS Safari, too.

Because Safari underpins all iOS browsers, Apple's technical decisions for its browser define what the nominally "open web" can and cannot offer developers and users. Critics argue that Apple uses its position to direct both developers and users to native apps, where the company collects a commission.

The best case study here is Safari's tepid adoption of WebGL, a JavaScript API designed to enable more complex browser-based 2D and 3D rendering using local processors. WebGL doesn't bring "app like" gaming to the browser, but it does elevate performance while also simplifying the development process.

However, Apple's mobile browser typically supports only a subselection of WebGL's total feature set, and often years after they're first released. Mac Safari adopted WebGL 2.0 18 months after it released, but mobile Safari waited more than four years to do the same.* In effect, Apple's iOS policies reduce the headroom afforded by already low ceilings of web-based gaming, thereby pushing more developers and users to its App Store, and avoiding an interoperable "Metaverse" that, like the World Wide Web, was built on HTML.

Support for this hypothesis can be found in the approach Apple has taken to another method of real-time rendering: cloud. In Chapter 6, I discussed this technology in detail; as you'll recall, cloud game streaming involves shifting much of the "work" normally managed by a local device (such as a console or tablet) to a remote data center. A user can then access computing resources that far outstrip those that might be affordably (if ever) contained in a small consumer electronics device, which is theoretically good for both user and developers.

It is not good, however, for those whose business models are predicated upon selling said devices and the software that runs on them. Why? These devices end up little more than a touchscreen with a data connection and that's simply playing a video file. If a 2018 iPhone

---

* That Apple now supports WebGL 2.0 is somewhat beside the point. Developers do not wait years in hopes of a given standard being supported and they cannot bet their futures.

and 2022 iPhone both play *Call of Duty* equivalently well—the most complex application that's likely to run on the device—why spend $1,500 on replacing the device? If you no longer need to download multi-gigabyte games, why buy the higher priced (and higher margin) iPhones with large hard drives?

Cloud gaming is even more threatening to Apple's relationship with mobile app developers. To release an iPhone game today, a developer must be distributed by Apple's App Store and use Apple's proprietary API collection, Metal. But to release a cloud-streaming game, a developer could distribute through nearly any application, from Facebook to Google, the *New York Times*, or Spotify. Not only that, but the developer could use whichever API collections they wanted, such as WebGL or even those the developer wrote themselves, while also using whichever GPUs and operating systems they liked—and still reach every Apple device that worked.

For years, Apple essentially blocked any form of cloud gaming application. Google's Stadia and Microsoft's Xbox were technically allowed to have *an* application, but only if it did not actually load games. Instead, they were effectively showrooms—showing off what these hypothetical services had—like a version of Netflix that had thumbnail tiles that could not be clicked.

Because cloud game streams are video streams, and the Safari browser supports video streaming, cloud gaming was still technically possible on iOS devices (though Apple prohibited these applications from telling users this fact). But Safari also places numerous experiential limitations on the Safari browser which, to both cloud- and WebGL-based game developers, make browser-based gaming unsatisfying. For example, web apps are not allowed to perform background data synchronization, automatically connect to Bluetooth devices, or send push notifications such as an invitation to play a game. Again, these limitations don't really affect applications like the *New York Times* or Spotify, but severely erode interactive ones.

Apple originally argued that cloud gaming was banned in order to protect users. Apple would not be able to review and approve all titles and their updates, and thus users could be harmed by inappropriate content, privacy violations, or substandard quality. But this argument was inconsistent with other app categories and policies. Netflix and YouTube bundle together thousands and even billions of videos that went unreviewed by Apple. In addition, Apple's App Store policy did not require developers to have perfect moderation, merely robust efforts and policies.

Given this, critics have countered that Apple's policies were motivated by the desire to protect its own hardware and game sale businesses. The rise of music streaming might have been a cautionary tale for Apple, in this regard. In 2012, iTunes had a nearly 70% market share in digital music revenues in the US and operated at nearly 30% gross profit margins. Today, Apple Music has less than a third of streaming music share and is believed to operate at negative gross margin. Spotify, the market leader, doesn't even sell itself through iTunes. Amazon Music Unlimited, which ranks third, is almost exclusively used by Prime customers, and nets Apple no revenue.

In the summer of 2020, Apple finally revised its policies so that services such as Google Stadia and Microsoft's xCloud could exist on iOS and as apps. But the new policies are byzantine and widely described as anti-consumer. To give just one striking example, cloud gaming services would need to first submit every single game (and future update) to the App Store for review, and then maintain a separate listing for the game in the App Store.

This policy requirement has several implications. First, Apple would effectively control the content release schedules for these services. Second, it could unilaterally deny any title (which would happen only after it had been licensed, and the service would have no direct ability to modify the game for it to meet Apple's requirements). Third, user reviews would be fragmented across the streaming service's app and the App Store. Fourth, these game-distribution services would need their developers to form a relationship with the App Store, a competing game-distribution service.

Apple's policies also stated that Stadia subscribers would still not be able to play Stadia games through the Stadia app (which would remain a catalogue). Instead, users would need to download a dedicated Stadia app for every individual game they wanted to play. This would be like downloading a *House of Cards* Netflix app, and an *Orange Is the New Black* Netflix app, and a *Bridgerton* Netflix app, with the Netflix app itself serving only as a catalogue/directory for rights management, rather than a streaming video service. According to leaked emails between Microsoft and Apple, each app would be nearly 150 megabytes, and need to be updated every time the underlying cloud-streaming technology was updated.

Even though Stadia would bill the user for their gaming subscription, curate the content inside that subscription, and power its delivery, Apple would distribute the cloud game (via the App Store) and iOS customers would access the title through the iOS home screen (not the Stadia app). Apple's policies also create inevitable consumer confusion. If a game was offered by multiple services, for example, the App Store would end up with multiple listings (there would be *Cyberpunk 2077*—Stadia, *Cyberpunk 2077*—Xbox, *Cyberpunk 2077*—PlayStation Now, and so on). And every time a service removed a title from their service (if Stadia removed *Cyberpunk 2077*), users would be left with an empty app on their device.

Apple also declared that all game streaming services would need to be sold through the App Store as well, treating them differently than how Apple treats other media bundles, such as those of Netflix and Spotify, which have their apps distributed by the App Store but can (and choose) not to offer iTunes billing. Finally, Apple said that every subscription-based game must also be made available as an à la carte purchase through the App Store. This, again, differs from its policies with music, video, audio, and books. Netflix does not need to (and doesn't) make *Stranger Things* available on iTunes for purchase or rental.

Microsoft and Facebook (which was also working on its own cloud game-streaming service) were quick to publicly criticize Apple's revised policy. "This remains a bad experience for customers," Microsoft reported the day of Apple's update. "Gamers want to jump directly into a game from their curated catalog within one app just like they do with movies or songs, and not be forced to download over 100 apps to play individual games [that stream] from the cloud." Facebook's video president of gaming told *The Verge*, "We've come to the same conclusion as others: web apps are the only option for streaming cloud games on iOS at the moment. As many have pointed out, Apple's policy to 'allow' cloud games on the App Store doesn't allow for much at all. Apple's requirement for each cloud game to have its own page, go through review, and appear in search listings defeats the purpose of cloud gaming. These roadblocks mean players are prevented from discovering new games, playing cross-device, and accessing high-quality games instantly in native iOS apps—even for those who aren't using the latest and most expensive devices."

## Blocking Blockchain

For all of the constraints that Apple places on interactive experiences, its most stringent controls focus on emergent payment rails.

Witness Apple's control over its NFC chip. NFC refers to near-field communication, a protocol that enables two electronic devices to wirelessly share information over short distances. Apple prohibits all iOS applications and browser-based experiences from using NFC mobile payments, with the sole exception being Apple Pay. Only Apple Pay can offer "tap-and-go" payments, which take a second or less to complete and don't even require the user to open their phone, let alone navigate to an application or its sub-menu. Visa, meanwhile, must ask a user to do exactly that, then have a retailer scan a virtually reproduced version of a physical card or a bar code.

Apple claims that its policies are intended to protect its customers and their data. But there's no evidence to suggest that Visa, Square, or Amazon would endanger users—and Apple could easily introduce a policy that provided NFC access only to regulated banking institutions. Alternatively, it could place additional security requirements, such as $100 or even a $5 limit, on NFC purchases. Apple does enable third-party developers to use the NFC chip for other use cases that are arguably more dangerous than buying a cup of coffee or a pair of jeans. Marriott and Ford, for example, use NFC to unlock hotel rooms and car doors. One might reasonably conclude this is correlated with the fact Apple doesn't operate in the hotel or automotive industries. It does, however, take an estimated 0.15% of every Apple Pay transaction—even if Apple Pay processes the actual transaction using the customer's Visa or MasterCard.

The Apple Pay problem may seem modest today. That said, and as I discussed in Chapter 9, we may be moving to a future in which our smartphone is not just a smartphone, but a supercomputer that will power the many devices around us. It's also likely to serve as our passport to the virtual and physical worlds. Not only is the Apple iCloud ID used to access most online software today, Apple has received approval from several American states to operate digital versions of state-issued identification, such as a driver's license, which can then be used to fill out a banking application or board a flight. Exactly how these IDs are used, which developers they're made available to, and under what conditions, could help determine the nature and timing of the Metaverse.

Another case study is Apple's approach to blockchains and cryptocurrencies. In the next chapter, I'll go into more detail on how these technologies work, what they might offer the Metaverse, and why Apple's policies are so problematic if you're a blockchain believer. But first, I want to quickly address how they're already in conflict with App Store policies and platform incentives. For example, neither Apple nor any of the major console platforms allow applications that are used for crypto mining or decentralized data processing. Apple has based this prohibition on the stated belief that such apps "rapidly drain battery, generate excessive heat, or put unnecessary strain on device resources."[11] Users might fairly argue that they—not Apple or Sony—have the right to decide whether their battery is being too quickly drained, to manage the health of their device, and to determine the appropriate use of their device's resources. Regardless, the net effect is that none of these devices can participate in the blockchain economy, nor make their idle computing power available to those who need it (via decentralized computing).

In addition, these platforms (with the exception of the Epic Games Store) do not allow games that accept cryptocurrencies as a form of payment, or that use cryptocurrency-based virtual goods (that is, non-fungible tokens, or NFTs). Though this is sometimes portrayed as a protest against the energy used to power blockchains, such claims don't hold up to scrutiny. Sony's music label has invested in NFT start-ups, and created its own NFTs, while Microsoft's Azure offers blockchain certifications and its corporate venture arm has made numerous start-up investments. Apple CEO Tim Cook has admitted that he owns cryptocurrencies and considers NFTs "interesting." It's more likely that these platforms refuse blockchain games because they simply do not work with their revenue models. Allowing *Call of Duty: Mobile* to connect to a cryptocurrency wallet would be akin to a user connecting the game directly to their bank account, rather than paying through the App Store. Accepting NFTs, meanwhile, would be like a movie theater permitting customers to bring their grocery bags to a film—some people might still buy a box of M&Ms, but most wouldn't. What's more, it's impossible to imagine how a platform might justify taking a 30% commission from the purchase or selling a multi-thousand- or million-dollar NFT—and if such commissions did apply, the entirety of the NFT's value would be devoured if it traded hands enough times.

Apple's efforts to support cryptocurrencies even as it protects its app store gaming revenue has produced more confusion. Apple enables users to buy and sell cryptocurrencies using trading applications such as Robinhood or Interactive Brokers, for example, but they cannot purchase NFTs through these same applications. What makes this distinction strange is the fact that there is no technical distinction between these two purchases—the only difference is that bitcoin is a "fungible" crypto-based token, in that every bitcoin is substitutable with another, while buying an NFT piece of artwork is a non-fungible token, in that it isn't substitutable with any other token. Things get more confusing if the right to this non-fungible token is fractionalized into fungible tokens (think of selling shares to a piece of artwork). These "shares" can be bought and sold via the iPhone app. Regardless, Apple's murky policies produce an experience that benefits neither developers nor customers—one which resembles that faced by cloud game-streaming apps. The iOS apps for NFT marketplaces such as OpenSea can only serve as a catalogue; users can see what they own and what others are selling—but to buy or trade themselves, they must move to the web browser. In addition, the only blockchain-based games that can run on the iPhone are those that use the web browser. This is why almost all of the hit blockchain games of 2020 and 2021 were focused on collecting (virtual sports cards, digital artwork, and so on) or otherwise limited to simple 2D graphics and turn-based play (*Axie Infinity*, for example, which is a sort of reimagining of the 1990s hit GameBoy game *Pokémon*). It's not possible to do much more.

## Digital First Requires the Physical First

At the core of the virtual payment rails problem is a conflict. The very idea of the Metaverse supposes that the "next platform" is not based on hardware, nor even an operating system. Instead, it is a persistent network of virtual simulations that exist irrespective of, and in fact, are agnostic regarding, a given device or system. The difference is that between a *New York Times* app that runs on a single user's iPhone and an iPhone used to access a living *New York Times* universe. There is evidence of this transition today. The most popular virtual worlds, such as those of *Fortnite*, *Roblox*, and *Minecraft*, are designed to run on as many devices and operating systems as possible, and are only lightly optimized for any specific one.

Of course, you can't access the Metaverse without hardware. And every hardware player is fighting to be a (if not *the*) payment gateway to this multi-trillion-dollar opportunity. To win this fight, they forcibly bundle their hardware with various APIs and SDKs, app stores, payment solutions, identities, and entitlement management, a process that increases store fees, staves off competition, and harms the rights of individual users and developers. We can see this through the blocking of WebGL, browser-based notifications, cloud gaming, NFC, and blockchains. There are always justifications for an individual policy, but they're impossible for the market to validate when there are only two smartphone platforms and their respective stacks are so extensively bundled. Even regulatory efforts to introduce more competition to individual service offerings have ended up stymied. In August 2021, a bill passed in South Korea banning app store operators from requiring their own payment systems, arguing such a requirement was monopolistic and harmed both consumers and developers. Three months later, and before the law changes were set to come into effect, Google announced that apps which chose to use an alternative payment service would have to pay a new fee for using their app store. Its price? Four percent less than the old fee—almost exactly the cost of its old fee, less the fees charged by Visa, MasterCard, or PayPal. As such, any developer who chose to use another payment rail would end up with savings of less than 1%. The margin was so small that changing systems would be senseless, and no price cut to consumers would be possible. In December 2021, Dutch regulators ordered Apple to let dating apps use third-party payment services (the category-specific requirements stemmed from the fact that category leader Match Group had filed a complaint with the Netherlands Authority for Consumers and Markets). In response, Apple updated its store policies in the Netherlands, allowing developers to issue (and thus maintain) a Dutch-only version of their app which supported alternative payments. However, this new version would not be able to use Apple's own payments solution, and Apple would enforce a new transaction fee of 27% (i.e., the old 30% minus 3%). Furthermore, the app would need to display a disclaimer that it would not "support the App Store's private and secure payment systems."[12] Various regulators, executives, and analysts argued the phrasing chosen by Apple was designed to "scare" users[13] and that developers would need to send Apple a monthly report detailing every single transaction placed under this system, after which they'd receive an invoice for the commissions owed (payable within 45 days).

The centrality and influence of hardware helps to explain why Facebook, in particular, is so committed to building its own AR and VR devices, and investing in fanciful projects such as brain-to-machine interfaces and smartwatches with their own wireless chips and cameras. As the only member of the big-tech giants without a leading device and/or operating system, Facebook is uniquely familiar with how operating solely on the platforms of its largest competitors is an impediment. Its cloud gaming service has effectively been blocked on every major mobile and console platform. And whenever Facebook sells something to one of its users, it collects as much net revenue as it sends to its own foes. The company's integrated virtual world platform, *Horizon Worlds*, meanwhile, is fundamentally constrained by the fact that it can never offer a developer a greater cut of revenues than iOS or Android. The most painful example may be Apple's "App Tracking Transparency" (ATT) changes, which were implemented in 2021, 14 years after the first iPhone. In a simplified sense, ATT required app developers to receive explicit "opt-in" permission from users in order to access key user and device data, while also explaining exactly what data was being collected and why (much of this script was written by Apple, and the company's App Store team would have approve rights over all alterations). Apple argued the changes were in the interest of users, 75% to 80% of whom were believed to have rejected the prompt by December 2021.[14] Others saw the move as a deliberate effort to stymie the company's advertising-focused competitors, build up Apple's own advertising business, and, by reducing the efficacy of advertising, prompt more developers to focus their business model on in-app payments, where Apple collected a 15%–30% fee. In February 2022, Mark Zuckerberg said Apple's policy change would reduce that year's revenue by $10 billion (roughly as much as Facebook was spending on its Metaverse investments). Some reports show that Apple's ad business was responsible for 17% of all iOS app installations before ATT was deployed. Six months later, it held nearly 60% market share.

To solve this problem, Facebook needs to do more than build its own low-cost, high-performance, and lightweight devices. It needs these devices to run independently of an iPhone or Android device—that is, without leveraging their computing or networking chips, as Apple and Google are likely to. The result is that Facebook's devices are likely to be more expensive, technically limited, and heavier than those produced by today's smartphone giants. This is, perhaps, why Mark Zuckerberg has said that "the hardest technology challenge of our time may be fitting a supercomputer into the frame of normal-looking glasses"—his competitors already put most of this supercomputer in a person's pocket.

For similar reasons, the most common pattern of disruption in the digital era—new computing devices—may be a false hope. The hegemony of Microsoft's Windows was broken by a stand-alone device, the mobile phone. But if our AR and VR headsets, smart lenses, and even brain-to-machine interfaces are governed by these same mobile phones, then there can be no new king.

## New Payment Rails

In this chapter, I've covered the role of payment rails in determining the "cost of doing business" in the digital era, and how they are influencing the technical, commercial, and competitive development of the Metaverse. What I've not directly addressed is how they can actively transform an economy. China provides a useful case study.

When Tencent's WeChat launched in 2011, China was primarily a cash society. But within the span of a few years, the messaging app hurled the country into the digital payments and services era. This was a consequence of many of WeChat's unique—and in the West, effectively impossible—opportunities and choices. For example, WeChat enabled users to connect directly to their bank account rather than require an intermediary credit card or digital payments network, which is prohibited by the major gaming consoles and smartphone app stores. Without intermediaries, and because Tencent wanted to build up its social messaging network, WeChat offered tiny transaction fees: 0%–0.1% for peer-to-peer transfers and less than 1% for merchant payments, with no fees for real-time delivery or payment confirmations. And because this payment capability was built upon common standards (QR codes) and built into a messaging app, it was easy for everyone with a smartphone to adopt and use. WeChat's success also helped Tencent build up the domestic video gaming industry, too, which would have otherwise been limited by the lack of credit cards across the country.

In the West, these systems would normally be at the mercy of the hardware gatekeepers. However, Tencent grew so powerful and so rapidly in China that even Apple was forced to allow WeChat to operate its own in-app app store, and directly process in-app payments—and iPhone launched in China two years before the messaging service. In 2021, WeChat processed an estimated US $500 billion in payments, with an average value of only a few dollars each.

For the Metaverse to emerge, it's likely that developers and creators in the West will need to find ways around the gatekeepers. Here, finally, we arrive at why there's such enthusiasm for blockchains.

# Chapter 11

# BLOCKCHAINS

SOME OBSERVERS TODAY BELIEVE THAT BLOCK-chain is structurally required for the Metaverse to become a reality, while others find that claim absurd.

There remains a good deal of confusion about blockchain technology itself, even before getting to its relevance to the Metaverse, so let's begin with a definition. Put simply, blockchains are databases managed by a decentralized network of "validators." Most databases today are centralized. A single record is kept in a digital warehouse, managed by a single company that tracks information. For example, JPMorgan Chase manages a database that tracks how much money you have in your checking account, as well as detailed records of prior transactions that validate how that balance was accumulated. Of course, JPMorgan has many backups of this record (and you might too), and it really operates a network of different databases, but what matters is that these digital records are managed and owned by a single party: JPMorgan. This model is used for almost all digital and virtual information, not just bank records.

Unlike a centralized database, blockchain records sit in no single location, nor are they managed by a single party—or, in many cases, even an identifiable group of individuals or companies. Instead, a blockchain "ledger" is maintained through consensus across a network of autonomous computers situated around the world. Each of these computers, in turn, is effectively competing (and being paid) to validate this ledger by solving what are essentially cryptographic equations that arise from an individual transaction. One benefit of this model is its relative incorruptibility. The larger (that is, the more decentralized) the network, the harder it is for any data to be overwritten or disputed as the majority of the decentralized network would have to agree, rather than, say, an individual at JPMorgan or the bank overall.

Decentralization has its downsides. For example, it is inherently more expensive and energy-consuming than using a standard database because so many different computers are performing the same "work." For similar reasons, many blockchain transactions take tens of seconds, or even longer, to complete as the network must first establish consensus—which can mean sending information across much of the world just to confirm a transaction two feet away. And of course, the more decentralized the network, the more challenging the problem of consensus typically becomes.

Due to the above issues, most blockchain-based experiences actually store as much "data" as they can in traditional databases, rather than "on chain." This would be like JPMorgan storing your account balance on a decentralized server, but your account log-in information and bank account in a central database. Critics argue that anything that is not fully decentralized is in effect fully centralized—in the above case, your funds are still effectively controlled and validated by JPMorgan.

This leads some people to contend that decentralized databases represent technical steps backward—less efficient, slower, and still dependent upon their centralized peers. And even if data is fully decentralized, the upside seems modest; few worry, after all, that JPMorgan and its centralized database might misplace its customers' account balances or steal from them. It's arguably scarier to think a collection of unknown validators are all that protect our wealth. If Nike said you owned a virtual sneaker, or managed and then tracked a record stating you sold it to another online collector, who would dispute it or discount its value because Nike was the one recording the transaction?

So why is a decentralized database or server architecture seen as the future? It helps to put aside the idea of NFTs, cryptocurrencies, fears of record theft, and the like. What matters is that blockchains are *pro-*

*grammable* payment rails. That is why many position them as the first digitally native payment rails, while contending that PayPal, Venmo, WeChat, and others are little more than facsimiles of legacy ones.

## Blockchains, Bitcoin, and Ethereum

The first mainstream blockchain, Bitcoin, was released in 2009. The sole focus of the Bitcoin blockchain is to operate its own cryptocurrency, bitcoin (the former is usually capitalized while the latter is not, in order to distinguish between the two). To this end, the Bitcoin blockchain is programmed to compensate processors handling bitcoin transactions by issuing them bitcoin (this is called a "gas" fee and is typically paid by the user to submit a transaction).

Of course, there's nothing novel about paying someone—or even many people—to process a transaction. In this case, however, the work and payment happen automatically and are united; a transaction cannot happen without the processor being compensated. This is part of why blockchains are referred to as being "trustless." No validator need wonder whether, how, and when they'll be paid, or if the terms of their payment might alter. The answers to these questions are transparently baked into the payment rail—there are no hidden fees, nor risks of sudden policy changes. Related, no user need worry about whether unnecessary data is being shared or stored by an individual network operator, or might then be misused. Contrast this with using a credit card stored on a centralized database that might later be hacked by an outside party or improperly accessed by an employee. Blockchains are also "permissionless": in the case of Bitcoin, anyone can become a network validator without needing to be invited or approved, and anyone can accept, buy, or use bitcoin.

These attributes create a self-sustaining system through which a blockchain can increase capacity while decreasing cost and improving security. As transaction fees increase in dollar value or volume, additional validators join the network, which decreases prices through competition. This, in turn, increases a blockchain's decentralization, which makes it harder for anyone trying to manipulate a ledger to establish consensus (think of an electoral candidate trying to tamper with 300 voting boxes versus three).

Advocates also like to highlight that the trustless and permissionless blockchain model means that the "revenue" and "profits" from operating its payment network are set by the market. This differs from the traditional financial services industry, which is controlled by a handful of decades-old giants with few competitors and no incentive to cut rates. The only competitive force on PayPal's fees, for example, are those charged by Venmo or Square's Cash App. For Bitcoin, fees are pushed down by anyone who chooses to compete for a transaction fee.

Not long after Bitcoin emerged (its creator remains anonymous), two early users, Vitalik Buterin and Gavin Wood, began developing a new blockchain, Ethereum, which they described as a "decentralised mining network and software development platform rolled into one."[1] Like Bitcoin, Ethereum pays those operating its network through its own cryptocurrency, Ether. However, Buterin and Wood also established a programming language (Solidity) that enabled developers to build their own permissionless and trustless applications (called "dapps," for decentralized apps), which could also issue their own cryptocurrency-like tokens to contributors.

Ethereum, then, is a decentralized network that is programmed to automatically compensate its operators. These operators do not need to sign a contract to receive this compensation, nor worry about being paid, and while they compete with one another for compensation, this competition enhances the performance of the network, which in turn attracts more usage, thereby producing more transactions to manage. In addition, with Ethereum, anyone can program their own applications on top of this network, while also programming this application to compensate its contributors, and, if successful, providing value to those who operate the underlying network, too. All of this occurs without a single decision-maker or managing institution. In fact, there is and can be no such body.

The decentralized governance approach does not prevent their underlying programming from being revised or improved. However, the community governs these changes and must therefore be convinced that any revisions are to their collective benefit.* Developers and users need not worry that, as an example, "Ethereum Corp" might suddenly increase Ethereum transaction fees or impose new ones, deny an emerging technology or standard, launch a first party service that competes with the most successful dapps, and so on. Ethereum's trustless and permissionless programming actually encourages developers to "compete" with its core functionality.

Ethereum has its detractors, who level three primary criticisms: its processing fees are too high, its processing times are too long, and its programming language is too difficult. Some entrepreneurs have chosen to address one or all of these problems by constructing competing blockchains, such as Solana and Avalanche. Other entrepreneurs instead built what are called "Layer 2" blockchains on top of Ethereum (the Layer 1). These Layer 2 blockchains effectively operate as "mini-blockchains," and use their own programming logic and network to manage a transaction. Some "Layer 2 scaling solutions" batch transactions together, rather than processing them individually. This naturally delays a payment or transfer, but real-time processing is not always required (just as your wireless phone-service provider doesn't need to be paid at a specific time of the day). Other "scaling solutions" look to simplify the process of transaction validation by polling just a portion of the network, rather than all of it. Another technique involves letting validators propose transactions without proving they've solved the underlying cryptographic equation, while keeping them honest by offering bounties to other validators if the latter prove this proposal as dishonest, with the bounty mostly paid by the dishonest valida-

---

* This is not automatically the case, as blockchains can be programmed to bestow (or withhold) a wide range of governance rights to token holders, while the creators of said blockchain control the initial distribution of these tokens. However, most major "public blockchains," in contrast to "private blockchains," which are typically owned by a corporation, are decentralized and community-run.

tor. These two approaches reduce the network's security, but many consider the trade-off appropriate for small-dollar-value purchases. Think of it as the difference between buying a coffee and buying a car; there's a reason why Starbucks doesn't require your credit card's billing address, while a Honda dealership does, along with a credit check and government ID. "Sidechains," meanwhile, allow tokens to be moved on and off of Ethereum as needed, serving a bit like a petty cash drawer versus a locked safe.

Some argue that Layer 2s are a patchwork solution—that developers and users would be better off working on higher performance Layer 1s. They might be right. Yet it's significant that a developer can use a Layer 1 to jumpstart their own blockchain, and then disintermediate that Layer 1 from its users, developers, and network operators by using, or even building, a Layer 2 blockchain. What's more, the trustless and permissionless programming of Layer 1s mean that competing Layer 1s can "bridge" into it, enabling developers and users to forever shift their tokens to another blockchain.

## The Arc of Android

An obvious contrast to trustless and permissionless blockchains are the policies of Apple and its iOS platform. However, iOS was never billed as an "open platform" nor as a community-centric one. In this regard, it's an unfair comparison. A better one would be with Android.

The Android OS was bought by Google for "at least $50 million" in 2005, and the search giant was always going to have an outsized role in its development. To assuage concerns, Google established the Open Handset Alliance in 2007, which would collectively steer the "open-source mobile operating system" based on open-source Linux OS Kernel, and would prioritize "open source technologies and standards." At launch, the OHA counted 34 members, including telecommunications giants China Mobile and T-Mobile, software developers Nuance Communications and eBay, component manufacturers Broadcom and

Nvidia, and device makers LG, HTC, Sony, Motorola, and Samsung. To join the OHA, members had to agree not to "fork" Android (take a copy of the "open-source" software and begin independently developing it) or support those who did (Amazon's Fire OS, which powers its Fire TV and tablets, is an Android fork).

The first Android released in 2008 and by 2012 the operating system had become the most popular in the world. The OHA and Android's "open" philosophy were less successful. In 2010, Google began building its own "Nexus" line of Android devices, which the company positioned as "reference devices" that would "serve as a beacon to show the industry what's possible."[2] Only a year later, Google purchased one of the largest independent manufacturers of Android devices, Motorola. In 2012, Google began moving its key services (maps, payment, notifications, the Google Play Store, and more) outside of the operating system itself and into a software layer, "Google Play Services." To access this suite, Android licensees would need to comply with Google's own "certifications." In addition, Google would not allow uncertified devices to use Android branding.

Many analysts considered Android's progressive closure a response to Samsung's growing success with the operating system. In 2012, the South Korean giant sold nearly 40% of Android-powered smartphones (and the majority of high-end ones)—more than seven times as many as the second-largest manufacturer, Huawei. In addition, Samsung had become increasingly aggressive with its alterations to the "stock" version of Android, producing and marketing its own interface (TouchWiz), while also preloading its devices with its own suite of apps, many of which competed with those offered by Google. Samsung even added its own mobile app store. Samsung's success as an Android manufacturer is inarguably connected to these investments, but their approach is not dissimilar from "forking" it. Regardless, Samsung's de facto TouchWiz OS threatened to disintermediate Google from its developers and users, while also serving as the true "reference device."

The arc of Android is important to any understanding of the future of the Metaverse. The Metaverse offers the opportunity to disrupt today's gatekeepers, such as Apple or Google, but many fear that we'll just end up with new ones—maybe Roblox Corporation, or Epic Games. While Tencent's WeChat has low fees for real world transactions, for example, the company has used its control over digital payments and video games to charge 40%–55% for all in-app downloads and virtual items—a sum that far exceeds that of Apple, whose power Tencent was able to overcome. Just as an entry on a blockchain ledger is considered incorruptible, many believe the blockchain itself is too.

# Dapps

Unlike the major blockchains, many dapps are only partially decentralized. The dapp's founding team tends to hold a large portion of the dapp's tokens (because they inherently believe the dapp will succeed, they have incentives to keep holding these tokens, too) and may therefore have the ability to alter the dapp at will. However, the success of a dapp depends on its ability to attract developers, network contributors, users, and often capital providers, too. This requires the sale and awarding of at least some tokens to outside groups and early adopters. And to maintain community support, many dapps make a commitment to what's called "progressive decentralization," which is sometimes explicitly programmed to be consistent with the trustless nature of blockchains.

This might seem like a conventional start-up approach. Most applications and platforms need to keep their developers and users happy—especially at launch. And over time, their creators (the founders and employees) see their equity stakes diluted. Perhaps they even go public, thereby making the app's governance "decentralized" and enabling anyone to permissionlessly become a shareholder. But this is where the nuances of the blockchain come into sharp focus.

As an application becomes more successful, it tends to become more controlling. Google's Android and Apple's iOS followed this path. Many technologists view the phenomenon as the natural arc of a for-profit technology business—as it accumulates users, developers, data, revenue, and so on, it uses its growing might to actively lock in developers and users. This is why it's difficult to export your account from Instagram and re-create it elsewhere. It's also why many applications close their APIs as they scale or face competition.

Facebook, for example, long allowed Tinder users to use their Facebook account as their Tinder profile. Tinder, of course, would rather its users have their own Tinder account—but Tinder isn't intended to be a lifelong service and it was more important, especially early on, that it prove easy to use. The application also benefited from allowing users to quickly place their "best" Facebook photos on the application, rather than being forced to dig through years of cloud storage. Facebook also allowed users to connect their social graph to Tinder, thereby enabling them to see if they had friends in common with a would-be match, and if so, who. Some users preferred matching with someone they could reference check, for safety reasons. Others liked being able to go on a date to make a true "first impression," and thus "swiped right" only on individuals with whom they had no friends in common. Although many Tinder (and Bumble) users enjoyed this social graph feature, Facebook shut it down in 2018—not long before it announced its own dating service, which was naturally based around its unique social graph and network.*

Most blockchains are structurally designed to prevent this arc. How? They effectively maintain what's valuable to a dapp developer—their tokens—while the user has custody of their data, identity, wallet, and assets (for example, their images), via records that are, again, on the

---

* Facebook still allows Tinder users to use their Facebook account to sign up and log in, and to populate their Tinder profile with photos from their Facebook profile. Keeping this functionality, while shutting down access to a user's social graph, does make sense. Facebook cannot stop users from repurposing photos uploaded to Facebook, as they're easy to save ("right click, save as") and, through "like counts," also helps a user identify their best photos, too. Furthermore, if Facebook users are going to use Tinder, Facebook benefits from knowing as much. At minimum, it enables Facebook to then recommend its dating service, which does still use its social graph, to this user.

blockchain. In a simplified sense, a fully blockchain-based Instagram would never store a user's photos, operate their account, or manage their likes or friend connections.* The service cannot dictate, let alone control, how this data is used. In fact, a competing service can launch and then immediately tap into this same data, thereby placing pressure on a market leader. This blockchain model does not mean applications are commodified—the real Instagram outmaneuvered its competitors in part because of its superior performance and technical construction—but we generally recognize that ownership of a user's account, social graph, and data to be the primary store of value.† By keeping most of this outside the hands of an application (or in this case, a dapp), blockchain enthusiasts believe they can disrupt the traditional developer arc.

We've arrived at a simplified understanding of blockchain operations, capabilities, and philosophies. But the technology remains well below modern expectations for performance (today, a blockchain-based Instagram would likely store almost everything off-chain and every photo would take a second or two to load). More importantly, history is littered with technologies that might have disrupted existing conventions, only to fall short of promise or potential. Might blockchains fare better?

## NFTs

The greatest indicator of what blockchains might accomplish is what they have already achieved. In 2021, total transaction value exceeded $16 trillion—over five times as much than digital payment giants Pay-

---

* In a simplified sense, this data is only "exposed" to the service on an as-needed basis.

† Some venture capitalists and technologists say that blockchains are "fat protocols" that support "thin applications," in contrast to the "thin protocol" and "fat application" model of today's internet. While the Internet Protocol Suite is enormously valuable—and thankfully, not a for-profit product—it does not operate a user's identity, store their data, or manage their social connections. Instead, all of this information is captured by those building on TCP/IP.

Pal, Venmo, Shopify, and Stripe combined. In the fourth quarter, Ethereum processed more than Visa, the world's largest payment network and 12th-largest company by market capitalization.

That this was possible without a central authority, managing partner, or even a headquarters—that it all happened via independent (and sometimes anonymous) contributors—is a marvel. What's more, these payments were made across dozens of different wallets (rather than limited to a tightly controlled network, as is the case with peer-to-peer rails such as Venmo or PayPal), could be made at any time (unlike ACH and wires), and were completed within seconds to minutes (unlike ACH). Both sender and receiver could confirm a successful or failed transaction (without an additional fee). In addition, none of these transactions required a user to have a bank account, nor did any businesses need to sign, let alone negotiate, a long-term agreement with any specific blockchains, blockchain processors, or wallet providers. And as we'll see, blockchain wallets can also be programmed for automatic debits, credits, reversals—and more.

Although the majority of this transaction volume reflected investments and trading in cryptocurrencies, rather than making payments, it was also backed by a wellspring of crypto-based development. The simplest productions are NFT collections. Developers and individual users will place the ownership of an item (say, an image) onto a blockchain, in a process called "minting," after which the right to the image is managed similarly to any cryptocurrency transaction. The difference is that the right is to a "non-fungible token," or a token that, unlike a bitcoin or a US dollar, which are fully substitutable with any other, is unique.

Blockchain advocates believe that this structure increases the value of these virtual goods because they provide the purchaser with a truer sense of "ownership." Consider the adage "possession is nine-tenths of the law."[3] Under centralized server models, a user can never truly take ownership of a virtual good. Instead, they are simply provided access to a good that's held, via digital record, on someone else's property (that is, a server). And even if the user took that data off that server and onto their own hard drive, that's not enough either. Why? Because the rest of the world needs to acknowledge that data and agree on its use. Blockchains can do this by design.

The sense of possession is augmented by another key property right: the unrestricted right to resale. When a user buys an NFT from a given game, the trustless and permissionless nature of a blockchain means that the game's maker cannot block the sale of that NFT at any point. They're not even actively informed of it (though the transaction is recorded on a public ledger). For related reasons, it is impossible for a developer to "lock" blockchain-based assets into their virtual world. If Game A sells an NFT, Games B, C, D, and so on can incorporate it if the owner so chooses—the blockchain ownership data is permission-less and the owner is in control of the token. Finally, token structures mean that even if a duplicate version of this virtual good is minted, the original remains distinct and "original"—like a signed and dated painting listed as one of one.

Throughout 2021, roughly $45 billion was spent on NFTs and across a wide variety of categories.[4] These included Dapper Labs' NBA Top Shots, which turned individual moments from the 2020–2021 and 2021–2022 NBA seasons into collectible, trading-card-like NFTs; Larva Labs' Cryptopunks, a series of 10,000, algorithmically generated 24 × 24–pixel 2D avatars that are typically used as profile pictures; Axies, which are a sort of blockchain-based Pokémon that can be collected, bred, traded, and battled; and 3D horses used on *Zed Run*'s virtual casino racetracks. Bored Apes, another profile picture NFT series, are also used as a form of membership card to the Bored Apes Yacht Club.

Forty-five billion dollars is enough for even virtual eyes to pop, but it's not clear exactly how one might compare this sum to the nearly $100 billion spent in 2021 on video game content managed by a tra-ditional database. If someone purchases a Cryptopunk for $100, then sells it for $200, a total of $300 has been "spent," but only $100 has been spent on a net basis. Conversely, almost all purchases for *tradi-tional* virtual goods are one-way—that is, the goods cannot be resold or traded. Every dollar out is "net." This means that in 2022, another $100 billion might be spent on traditional game assets, but even if NFT spending doubles, there might only be $10 billion or so spent incrementally. Suddenly, the argument NFTs generated half the revenue of the game industry seems to have been exaggerated by a factor of ten. Perhaps a more accurate contrast would be between each year spend on traditional virtual assets and the market value of NFTs. The floor market cap for the 100 largest NFT collections was estimated at roughly $20 billion by the end of 2021—roughly half the trading volume, but still a quarter of the traditional gaming market. However, "floor market caps" assume every NFT in a given collection would be sold at the price of the lowest-priced NFT in that collection. This sort of analysis is a helpful way to compare the growth in different collections, but not their market value.

Some critics argue that most of the value in NFTs is speculative— i.e., based on the potential of profit—not based on utility, as is the case with *Fortnite* skins. This would make any sort of comparison impossible. At the same time, the global art market recognized $50.1 billion in spend (from buying and trading) in 2021, and few would debate the underlying purchases lacked utility, even though they also have speculative value. The closeness between these two categories is also instructive as to the scale of the NFT market. Furthermore, it's the very fact that NFTs can be resold that blockchain enthusiasts believe users place more value on them. NFTs can even be lent to other players or games, with the owner receiving a programmatic "rent" as these NFTs are used or "yield" when they generate revenue.

Irrespective of whether one should, or how they might, compare NFT spending to that of video game items and content, their growth rates are starkly different—as is their foreseeable growth potential. Overall spending on NFTs in 2021 was more than 90 times that of the roughly $350 million to $500 million spent on NFTs a year earlier, which in turn was more than five times that of 2019. In contrast, sales of traditional virtual items grew at a roughly 15% compound average rate. In addition, the utility of NFTs is severely constrained today by the fact that most video games do not yet support them. And because none of the major console platforms or mobile app stores support purchasing in blockchain-based games, most of the games that do use NFT titles are limited to the web browser and as a result have rudimentary graphics and gameplay. This is one of the reasons why many of the most successful NFT experiences are based around collecting, rather than active "play." It is also why the majority of the most popular games, game franchises, media franchises, brands, or companies haven't even issued NFTs—and why only a few million people are believed to have purchased an NFT, whereas billions of people make in-game purchases each year. As the functionality of NFTs improves, and the number of brands and participating users increases, the value of NFTs will of course grow. There's certainly a lot of headroom to each.

The most important upside may come from realizing interoperability in NFTs. While members of the blockchain community often say that blockchain NFTs are inherently interoperable, this isn't really true. I have mentioned that using a virtual good requires both access to its data as well as code to understand it. Most blockchain experiences and games do not have such code. In fact, most NFTs today place the rights to the virtual good on the blockchain, but not the virtual good's data, which remain stored on a centralized server. As such, the NFT's owner cannot export the good's data to another experience unless it receive permission from the centralized server that stores it. For similar reasons, almost no blockchain-based experiences are truly decentralized—even those that issue NFTs. The developers may not, for example, be able to revoke the rights to these NFTs but they could alter the code that uses it, or delete a user's in-game account.

The fact that "decentralized" assets have "centralized" dependencies leads to two major conclusions. First, NFTs are useless—propped up by fraud, speculation, and misunderstanding. This was often the case in 2021 and is likely to remain largely true for years to come. Second, the untapped potential of this technology is extraordinary and will be realized as the utility of, and access to, blockchain-based games and products expands.

This second conclusion points to the importance of blockchain for the Metaverse. For example, blockchains don't just establish a common and independent registry for virtual goods; they also provide a potential technical solution for the biggest obstacle to virtual goods' interoperability: revenue leakage.

Many players would love to bring their assets and entitlements from game to game. However, a number of game developers generate the bulk of their revenue by selling players goods that are exclusively used inside their games. The ability for a player to "buy elsewhere, use here" endangers a game developer's business model. Players might accumulate so many virtual goods that they no longer see the need to buy any more. Alternatively, players might start buying all of their skins from Game A, but then exclusively play them in Game B, which would result in distortions of where the majority of costs and revenues occur. In fact, it's likely that virtual goods sellers would emerge that could deeply underprice the goods sold inside the game because they don't need to recoup on a game's initial development nor operating costs.

Many developers are held back by the worry that an open-item economy might create far more value than they themselves capture. Developer A might produce Skin A for Game A, only for Game A to decline, and Skin A becomes a popular (and valuable) item in Developer B's longer running title. In this case, Developer A has actually created content for a competitor that beat them! Or maybe it just turns out that Developer A's creations have become iconic and highly valuable, thereby allowing a player to make far more profit from Developer A's creations than Developer A ever might. (Making matters worse, Developer A might never see an additional dollar after the initial sale.)

Trade is, of course, a messy process that involves some losers, even if the aggregate economic impact is strongly positive. However, interoperability can be partly facilitated with a mixture of taxes and duties (as is the case in the real world). For example, most NFTs are programmed to automatically pay its original creator a commission upon trade or resale. Similar systems can be established to pay upon importing or using a "foreign" good. Other observers propose programmed degradation of virtual goods, thereby attaching an implied "cost" to "use" that slowly removes value from a good and drives repurchasing. Blockchain programming cannot alone stop leakage, as prevention requires these systems and incentives to be "perfect"; the lessons of globalization tell us this is impossible. But through its trustless, permissionless, and automatic compensation models, many believe blockchains can nevertheless produce a more interoperable virtual world.

## Gaming on the Blockchain

Regardless of one's long-term belief in NFTs, there are more interesting aspects of blockchain-based virtual worlds and communities. Earlier on, I noted that dapps could issue their own cryptocurrency-like tokens to their network and users. These need not be issued for computing resources, as is the case with transaction processing of bitcoin and Ethereum. They can also be awarded for contributing time, delivering new users (customer acquisition), data entry, IP rights, capital (money), bandwidth, good behavior (such as community scores), helping to moderate, and more. These tokens can be provided with governance rights and, of course, may appreciate in value alongside the underlying project. Every user (that is, player) can often buy these tokens, too, enabling them to participate in the financial success of the games they love.

Developers believe that this model can be used to reduce the need for investor funding, deepen their relationship with the community, and significantly increase engagement. If we love to play *Fortnite*, or use Instagram, it stands to reason that we'll invest in and use them more if we can profit from and/or help govern them. After all, millions of people spent billions of hours tilling fields and sowing crops in *Farmville* for neither income nor ownership of *Farmville*, or even their own farms. As is always the case, blockchains are not a technical requirement for these sorts of experiences, but many believe its trustless, permissionless, and frictionless structures make such experiences more likely to take off, thrive, and, most importantly, prove sustainable. Sustainability stems not just from increased user involvement in and ownership of an application, but from the ways blockchain discourages the application from betraying user trust and instead forces the application to earn it.

A good example of dapp-to-user blockchain dynamic is demonstrated through the competition between Uniswap and Sushiswap. Uniswap was one of the first Ethereum dapps to gain mass adoption, having pioneered the automated market maker model, which allowed users to swap one token for another through a centralized exchange. Uniswap's predominantly open code was copied and forked by a competitor, Sushiswap. To gain adoption, Sushiswap issued tokens to its users. Users had the exact same functionality as they had from Uniswap, but received what was effectively an equity stake in Sushiswap for doing so. This forced Uniswap to counter by offering its own token, while retroactively rewarding all prior users. A user-beneficial "arms race" like this is typical. Dapps have few barriers that prevent the emergence of better versions of their functionality, specifically because blockchains, not dapps, maintain much of the data we typically value in the digital era—a customer's identity, data, and digital possessions, etc.

In addition to operating dapps and account services, blockchains can also be used to support the provision of compute-related gaming infrastructure. In Chapter 6, I highlighted the insatiable need for more computing resources and the long-held belief that realizing the Metaverse would require tapping into the billions of CPUs and GPUs that sit mostly unused at any given point in time. Several blockchain-based startups are pursuing this—and they are succeeding. One, Otoy, created the Ethereum-based RNDR network and token so that those who needed extra GPU power could send their tasks to idle computers connected to the RNDR network, rather than to pricey cloud providers such as Amazon or Google. All of the negotiation and contracting between parties is handled within seconds by RNDR's protocol, neither side knows the identity or specifics of the task being performed, and all transactions occur using RNDR cryptocurrency tokens.

Another example is Helium, which the *New York Times* has described as "a decentralized wireless network for 'internet of things' devices, powered by cryptocurrency."[5] Helium works through the use of $500 hotspot devices which allow their owner to securely rebroadcast their home internet connection—and up to 200 times faster than a traditional home Wi-Fi device. This internet service can be used by anyone, from consumers (say, to check Facebook) to infrastructure (e.g., a parking meter processing a credit card transaction). Transportation company Lime is a top customer and uses Helium to track its fleet of more than 100,000 bikes, scooters, mopeds, and cars, many of which regularly encounter mobile network "deadzones."[6] Those operating a Helium hotspot are compensated with Helium's HNT token, and in proportion to usage. As of March 5, 2022, Helium's network spanned more than 625,000 hot spots, up from fewer than 25,000 roughly a year earlier, distributed across nearly 50,000 cities in 165 countries.[7] The total value of Helium's tokens exceeds $5 billion.[8] Notably, the company was founded in 2013, but struggled to gain adoption until it pivoted from a traditional (i.e., unpaid) peer-to-peer model to one which offered contributors direct compensation via cryptocurrency. The long-term viability and potential of Helium remains uncertain; most internet service providers (ISPs) prohibit their customers from rebroadcasting their internet connection, and while the ISPs typically have ignored such service violations as long as the connection was not resold and total data usage was low, there is no guarantee that the ISPs will continue to ignore such violations by users of Helium or any analogous system. Regardless, the company serves as another reminder of the potential in decentralized payment models, and is now striking deals directly with ISPs.

The scale and diversity of the crypto-gaming boom in 2021, matched with its relative infancy and enormous revenues per player, have led to a surge in development. One of the leading gaming investors in the world told me that nearly every talented game developer she knew, with the exception of those already running world famous studios, was focused on building games on the blockchain. In total, blockchain-based games and gaming platforms received more than $4 billion[9] in venture investment (total VC funding for blockchain companies and projects was roughly $30 billion; some speculate another $100 billion–$200 billion more has already been raised or earmarked by venture funds).[10]

The influx of talent, investment, and experimentation can quickly produce a virtuous cycle whereby more users set up a crypto wallet, play blockchain games, and buy NFTs, increasing the value and utility of all other blockchain products, which also attracts more developers, and in turn more users, and so on. Eventually, this leads us to a future in which a handful of exchangeable cryptocurrencies are used to power the economies of countless different games, replacing one where spending remains fragmented across Minecoins, V-Bucks, Robux, and countless other proprietary denominations. And in this future all virtual goods are at least partly intended for interoperation.

At enough scale, even the most successful game developers of the pre-blockchain era, including Activision Blizzard, Ubisoft, and Electronic Arts, will find the technologies financially irresistible and competitively essential. The transition will be eased by the fact that they'll be opening up their economies and account systems to a system that is owned not by their platform competitors, such as Valve and Epic Games, but by the gaming community.

## Decentralized Autonomous Organizations

The most disruptive aspect of digitally native "programmable" payment rails, however, is how they enable greater independent collaboration and easier funding of new projects. This is not a structurally separate point from anything I've discussed thus far, but it's important to understand it in a broader context.

To this end, I want to talk about a vending machine. The first of these devices actually emerged millennia ago (around AD 50) and allowed a consumer to insert a coin and receive holy water in return. By the late

1800s, these machines supported a wide variety of different purchases—not just a single item, such as water, but also gum, cigarettes, and postage stamps. No shopkeeper or lawyer managed the distribution of goods, nor accepted and validated payment, but the system worked through fixed rules: "if this, then this." Everyone trusted the system.

Blockchains can be thought of as a virtual vending machine. Only much, much smarter. For example, they can track multiple contributors and value them differently. Imagine someone wanted to buy a candy bar from a real-world vending machine. Maybe she only had $0.75 and wanted to purchase a $1 candy bar, so she asked a passerby for 25¢ to complete the transaction. Perhaps they agreed, but only if they received half of the candy bar, rather than their pro rata share of a quarter. A "blockchain vending machine" would enable the two collaborators to write what's called a "smart contract" for this arrangement, and then after accepting each individual payment, the device would then automatically (and incorruptibly) deliver the appropriate amounts (half and half) to the appropriate owner. At the same time, the blockchain vending machine might have automatically paid everyone responsible for that candy bar as well—5¢ to the person who stocked the machine, 7¢ to the machine's owner, and 2¢ to the manufacturer.

Smart contracts can be written in minutes and serve almost any purpose; they can be small and temporary, or massive and persistent. A number of independent authors and journalists use smart contracts to fundraise for their research, investigations, and writing—serving as a sort of advance against future earnings, but one that comes from the community rather than a corporation. Upon completion, their works are minted to the blockchain and sold, or perhaps put behind a crypto-based paywall, with the proceeds shared back to their patrons. In other cases, a collective of authors have issued tokens to fundraise for a new, ongoing magazine that is then exclusively available to token-holders. Some writers use smart contracts to automatically share tips with those who helped or inspired them. None of this requires credit card numbers, entering ACH details, invoices, or even, really, much time—just a cryptowallet with cryptocurrency.

Some envision smart contracts as the Metaverse-era version of the

LLC (limited liability corporation) or 501(c)(3) (nonprofit organization). A smart contract can be written and instantaneously funded, with no need for participants to sign documents, perform credit checks, confirm payments or assign bank account access, hire lawyers, or even know the identities of the other participants. What's more, the smart contract "trustlessly" manages much of the administrative work for the organization on an ongoing basis, including the assignment of ownership rights, calculation of votes on bylaws, distribution of payments, and so on. These organizations are typically called "Decentralized Autonomous Organizations," or "DAOs."

In fact, many of the most expensive NFTs have been purchased not by individuals, but by DAOs comprising dozens (and in some cases, many thousands) of pseudonymous crypto users who could never have made the purchase on their own. Using the DAO's tokens, the collective can determine when these NFTs sell and at which minimum price, while also managing disbursements. The most notable example of such a DAO is the ConstitutionDAO, which was formed on November 11, 2021, to purchase one of the thirteen surviving first editions of the United States Constitution, which was to be auctioned by Sotheby's on November 18. Despite limited planning and no "traditional" bank account, the DAO was able to raise more than $47 million—far more than the $15 million–$20 million that Sotheby's estimated would be needed to win the auction. ConstitutionDAO ultimately lost to a private bidder, the billionaire hedge fund manager Ken Griffin, but *Bloomberg,* reporting on the effort, wrote that it "showed the power of the DAO . . . [DAOs have] the potential to change the way people buy things, build companies, share resources and run nonprofits."[11]

At the same time, ConstitutionDAO also illuminated many of the problems with the Ethereum blockchain. For example, an estimated $1 million to $1.2 million was spent processing transactions to fund the DAO. Though this represented 2.1% of contributions—within the average range for traditional payment rails—the median contribution was estimated at $217, with nearly $50 spent in "gas." In addition, the Ethereum blockchain cannot "waive" fees for reversing or refunding a transaction. As a result, these fees were effectively doubled as a result of the auction, as most contributors reclaimed their donations. Many donations remain in the DAO because the cost to retrieve a contribution exceeds its value. (Many of these problems are attributed to sloppy smart contract coding and could have been avoided, especially if another blockchain or Layer 2 solution had been used.)

Though a member of "traditional finance" was able to outmaneuver the "decentralized finance" community for the US Constitution, the world of high finance is also using DAOs to make their investments. One such example is the Komorebi Collective, which makes venture investments into "exceptional female and nonbinary crypto founders," and includes among its members a number of high-profile venture capitalists, technology executives, journalists, and human rights workers. In late 2021, some 5,000 outdoor enthusiasts used a DAO to purchase a 40-acre plot of land near Yellowstone National Park in Wyoming, which had passed legislation recognizing the legitimacy of DAOs earlier in the year. "CityDAO" is mostly organized through Discord and has no official leader (Ethereum co-founder Vitalik Buterin is a member), with all major decisions made through vote and members able to sell their membership tokens at any time. One member, CityDAO's de facto figurehead, told *Financial Times* that he hoped Wyoming's embrace of the DAO structure would "become this fundamental link between digital assets, crypto and the physical world."[12] As a point of reference, Wyoming was also the first state to authorize the creation of LLCs, having passed related legislation in 1977, some 19 years before it was available nationwide.

Friends with Benefits is effectively a DAO-based membership club where tokens are used to gain access to private Discord channels, events, and information. Some have argued that by requiring users to buy tokens to gain entry, FWB is simply replicating the centuries-old "membership dues" model of every exclusive club that came before it only now benefiting from the "crypto" hype. However, this view ignores the potency of FWB's token design. Members do not pay annual "dues." Instead, they need to buy a certain number of FWB tokens to gain entry—and then hold them to remain members. As a result, every member is a part owner of FWB, and can leave at any time by selling their tokens. Because these tokens appreciate as the club becomes more successful or desirable, every member is incentivized to invest their time, ideas, and resources into the club. Appreciation also makes it increasingly impractical for spammers to join FWB, whereas under normal circumstances the popularity of an online social platform only encourages trolls. Appreciation means the club must work harder to earn its ongoing role in a member's life. If you joined a club by buying $1,000 in tokens, but these tokens quadruple in value, the club must do more to keep you as a member. After all, if you leave, your sale depresses the market value of the remaining tokens. Finally, many social DAOs use smart contracts to issue tokens to individual members for their contributions, or to those who can't afford to join the collective but are deemed worthy by its members.

Nouns DAO is effectively a remix of FWB with Cryptopunks. Each day, one new Noun—an NFT of a cute pixelized avatar—is auctioned, with 100% of the net proceeds going into the Nouns DAO treasury, which exists exclusively to enhance the value of Nouns NFTs. How, specifically, does this treasury do so? By funding proposals authored by and voted on by owners of the NFTs. In effect, it is a constantly growing investment fund governed by a constantly growing board of governors.

Some see social DAOs and tokens as a way to address targeted harassment and toxicity on large-scale online social networks. Imagine, for example, a model whereby Twitter users were awarded valuable Twitter tokens for reporting poor behavior, could earn more for reviewing previously reported tweets, and lost them if they violated the rules. At the same time, rather than rely on tips or posting promotional tweets on behalf of advertisers to generate income, super-users and influencers could be awarded tokens for hosting events. By the end of 2021, Kickstarter, Reddit, and Discord had all publicly described plans to shift to blockchain-based token models.

# Blockchain Obstacles

There are still numerous obstacles facing a potential blockchain revolution. Most notably, blockchain remains too expensive and slow. For this reason the majority of "blockchain games" and "blockchain experiences" are still running mostly on non-blockchain databases. As a result, they are not truly decentralized.

Given the computational requirements of large-scale real-time rendered 3D virtual worlds, as well as their need for ultra-low latency, some experts debate whether we can ever fully decentralize such an experience—let alone "the Metaverse." Put another way, if computing is scarce and the speed of light already a challenge, how could it ever make sense to perform the same "work" countless times and wait for a global network to agree on the right answer? And, even if we could pull this off, wouldn't the energy use melt the planet?

This may sound too glib, but opinions vary. Many people believe the key technical problems will be resolved in time. Ethereum, for example, continues to overhaul its validation process so that network participants can perform less work (and crucially, less duplicative work), and it already uses less than a tenth of the energy per transaction of the Bitcoin blockchain. Layer 2s and sidechains are also proliferating, solving many of the shortfalls of Ethereum, while newer Layer 1s, such as Solana, are matching its programming flexibility but with far better performance. The Solana Foundation claims that a single transaction uses about as much energy as two Google searches.

In most countries and US states, DAOs and smart contracts are not legally recognized. This is beginning to change, but legal recognition is not a complete solution. There is a common adage: "the blockchain doesn't lie," or "the blockchain can't lie." That may be true, but users can lie to the blockchain. A musician might tokenize the royalties to their song, thereby ensuring smart contracts execute all payments. However, those royalties may not be received "on chain." Instead, a music label might send a wire to that musician's centralized database, and then the musician must put the appropriate sums into the appropriate wallet, and so on. And many NFTs are minted by those who don't own the rights to the underlying works. Blockchains, in other words, do not make everything trustless—just as contracts don't solve for all bad behavior.

Then there's the app store issue: if Apple and Google don't allow blockchain-games or transactions, what's the point? Well, blockchain maximalists believe that the totality of its economic forces will force even the world's mightiest corporations to change, rather than just game makers and gaming conventions.

## How to Think about Blockchains and the Metaverse

There are, as I see it, five ways to think about the blockchain's significance, both within the context of the Metaverse and within society at large. First, it's a wasteful technology propped up by scams and fads, and it receives attention not because of its merits, but due to short-term speculation.

Second, blockchains are indeed inferior to most, if not all, alternative databases, contracts, and computing structures, but may nevertheless lead to cultural change around user and developer rights, interoperability in virtual worlds, and compensation for those who support open-source software. Perhaps these outcomes were already inevitable, but blockchains may usher them in more quickly, and democratically.

Third, and more hopefully, blockchains will not become the dominant means for storing data, computing, payments, LLCs and 501(c)(3)s, and so on, but they will become key to many experiences, applications, and business models. Nvidia's Jensen Huang has argued that "blockchains [are] going to be here for a long time and [will] be a fundamental new form of computing,"[13] while global payment giant Visa has launched a cryptocurrency payment division, with its landing page declaring "Crypto is reaching extraordinary levels of adoption and investment—opening a world of possibilities for businesses, governments and consumers."[14] Recall from Chapter 8 the many problems which arise when one virtual world wants to "share" a unique asset with another, as would be the case with using an avatar bought in Epic Games' *Fortnite* but inside Activision's *Call of Duty*. Where is the asset stored when it's not in use: Epic's server, Activision's server, both, or somewhere else altogether? How is the storer compensated? If the item is altered or sold, who manages the right to make such a change and record it? How do these solutions scale to hundreds, if not billions, of different virtual worlds? If all blockchains do is offer an independent system which partly addresses some of these problems, many believe it will still produce a revolution in virtual culture, commerce, and rights.

A fourth view holds that blockchains are not just critical technologies for the future but also the key to disrupting today's platform paradigms. Recall why closed platforms tend to win. Free, open-source, and community-run technologies have been available for decades, and often promise developers and users a more fair and prosperous future, only to lose out to paid, closed, and privately owned alternatives. This is because the companies that operate these alternatives can afford enormous investments in competing services and tools, engineering talent, customer acquisition (for example, below cost hardware), and exclusive content. Such investments, in turn, attract users, producing a lucrative market for developers, and/or they attract developers, thereby attracting users who bring additional developers. Over time, the corporation that manages these developers and users leverages that control, alongside their ever-growing pool of profits, to lock in those same groups and stymie competitors.

How might blockchains alter this dynamic? They provide a mechanism through which significant and diverse resources—from wealth to infrastructure and time—can be easily aggregated and at a scale that contends with the mightiest of private companies. In other words, the only way to fight trillion-dollar corporate giants pursuing trillion-dollar opportunities is billions of people contributing trillions more.

Blockchains also have a baked-in economic model to compensate those who contribute to its success or ongoing operations, rather than rely on altruism and empathy, as is the case with most open-source projects. Moreover, blockchain-based experiences seem, at least thus far, to promise developers far greater profits than closed gaming platforms do. Just as important, the leaders of blockchain platforms and companies have significantly less control over their users and developers than those who build on traditional databases and systems, in that they cannot forcibly bundle a user's identity, her data, payments, content, services, and so on. Chris Dixon, a crypto-focused venture capitalist at Andreessen Horowitz, argues that if the dominant ethos of Web 2.0 was "Don't be evil," the phrase that (in)famously served as Google's unofficial motto, then a (blockchain-based) Web3 is "Can't be evil."

It's unlikely, however, that all data is "on chain," meaning few experiences will be fully "decentralized" and therefore remain de facto centralized or at least strongly controlled by a given party. In addition, control does not just come from data ownership, but from proprietary code and IP. It's relatively easy to copy Uniswap's code, which is mostly open-source, but the ability to copy the code that runs a blockchain-based *Call of Duty*'s doesn't mean a developer has the right to do so. A Disney blockchain game may provide users with indefinite rights to Disney-based NFTs, but that doesn't mean other developers can build Disney games with Disney's IPs. Put another way, a child can tell their own stories in the bathtub while using a Darth Vader action figure and a Mickey Mouse figurine, but Hasbro can't buy these figures and use them to sell a Disneyland board game. Another form of "lock in" is habit—the search results offered by Bing may be more accurate (and less ad-laden) than those of Google, but few of us think to use it. And even if they are better, how much better do they need to be to convince a user to change behaviors or overcome the synergies of using Google's search engine and browser? While Dixon's point is exaggerated, you'll note the examples above speak to how independent developers and creators establish power—rather than the ways in which the underlying platform (for example, Ethereum) builds or protects its own.

In general, society believes the rights of the former group are more important to economic health than those of the latter.

The fifth perspective on blockchains suggests they are essentially a requirement for the Metaverse—at least one which meets our lofty imaginations and we would actually want to live in. In 2017, Tim Sweeney said that we will "come to the realization that the blockchain is really a general mechanism for running programs, storing data, and verifiably carrying out transactions. It's a superset of everything that exists in computing. We'll eventually come to look at it as a computer that's distributed and runs a billion times faster than the computer we have on our desktops, because it's the combination of everyone's computer."[15] If we ever hope to produce richly, real-time rendered, and persistent world simulations, figuring out how to leverage the world's entire supply of computing, storage, and networking infrastructure will be necessary (though this doesn't require blockchain technology).

In January 2021, not long before the public craze over the Metaverse and NFTs began, Sweeney tweeted, "blockchain based underpinnings for an open Metaverse. This is the most plausible path towards an ultimate long term open framework where everyone's in control of their own presence, free of gatekeeping." In a follow-up tweet, Sweeney added two disclaimers: "1) The state of the art is far from the 60Hz transactional medium needed for 100M's of concurrent users in a real-time 3D simulation" and "2) Don't read this as an endorsement of cryptocurrency investment; that's a wild, speculative mess . . . But the tech is going places."[16]

In September 2021, Sweeney remained optimistic about the potential in blockchain, but also seemed discouraged by its misuse, declaring "[Epic Games isn't] touching NFTs as the whole field is currently tangled up with an intractable mix of scams, interesting decentralized tech foundations, and scams."[17] The following month, Steam banned games that used blockchain technology, prompting Sweeney to announce that "Epic Games Store will welcome games that make use of blockchain tech provided they follow the relevant laws, disclose their terms, and are age-rated by an appropriate group. Though Epic's not using crypto in our games, we welcome innovation in the areas of technology and finance."[18] Sweeney's critiques highlighted a problem often overlooked by blockchain enthusiasts, a group which typically sees decentralization as only a way to protect wealth, rather than also a way to lose it. Without intermediaries, regulatory oversight, or identity verification, the crypto space has become rampant with copyright violations, money laundering, theft, and lies. Many NFTs and blockchain-based games are propped up by user confusion around what exactly is being bought, how it can be used, and how it might be in the future (many don't care as long as the prices go up).

How much of the blockchain remains hype versus how much is (potential) reality remains uncertain—not unlike the current state of the Metaverse. However, one of the central lessons of the computing era is that the platforms that best serve developers and users will win. Blockchains have a long way to go, but many see their immutability and transparency as the best way to ensure the interests of these two constituencies remain prioritized as the Metaverse economy grows.

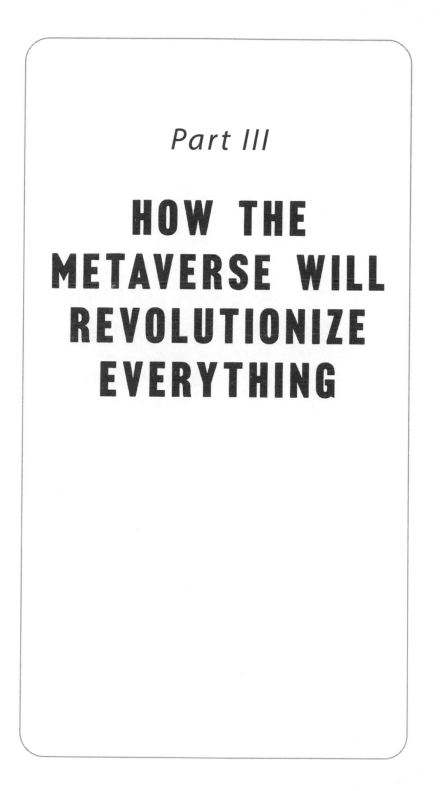

*Part III*

# HOW THE METAVERSE WILL REVOLUTIONIZE EVERYTHING

## Chapter 12

# WHEN WILL THE METAVERSE ARRIVE?

IN PART II, I OUTLINED WHAT'S REQUIRED TO realize the full vision of the Metaverse, as I've defined it. This first chapter of Part III takes up the inevitable question that follows—When will the Metaverse arrive?—and predicts what arrival will look like across a range of industries.

Even those pouring tens of billions per year into the "quasi-successor state" to the internet tend to disagree on the timing of the Metaverse's emergence. Satya Nadella, Microsoft's CEO, has said that the Metaverse is "already here," with Microsoft's founder Bill Gates forecasting that in "the next two or three years, I predict most virtual meetings will move from 2D camera image grids to the metaverse."[1] Facebook's CEO Mark Zuckerberg has said a "lot of [it] will become mainstream in the next five to 10 years,"[2] while Oculus's former and now consulting CTO John Carmack usually predicts an even later emergence. Epic's CEO Tim Sweeney and Nvidia's CEO Jensen Huang tend to avoid a specific timeline, instead saying the Metaverse will emerge over the coming decades. Google's CEO Sundar Pichai merely says that immersive computing is "the future." Steven Ma, the Tencent senior vice president who runs most of the company's gaming business and publicly introduced the company's "hyper digital reality" vision in May 2021, cautions that while "the metaverse's day will come[,] that day is just not today. . . . What we see today is indeed a leap from what we had just a few years ago. But it's also still primitive [and] experimental."[3]

To predict the future of the internet and computing, it helps to review their intertwined past. Ask yourself: When did the mobile internet era begin? Some of us might date this history from the very first mobile phones. Others might point to the commercial deployment of 2G, the first digital wireless network. Perhaps it really began with the introduction of the Wireless Application Protocol standard in 1999, which gave us WAP browsers and the ability to access a (rather primitive) version of most websites from nearly any "dumbphone." Or maybe the mobile internet era started with the BlackBerry 6000, or 7000, or 8000 series? At least one of them was the first mainstream mobile device designed for on-the-go wireless data. Most people, however, would likely say that the answer is linked to the iPhone, which arrived almost a decade after WAP and the first BlackBerry, nearly two decades after 2G, and 34 years after the first mobile phone call. It has since defined many of the mobile internet era's visual design principles, economics, and business practices.

In truth, however, there's never a moment when a switch flips. We can identify when a specific technology was created, tested, or deployed, but not when an era precisely began, or ended. Transformation is an iterative process in which many different changes converge.

Consider, as a case study, the process of electrification, which began in the late 19th century and ran midway into the 20th century, and focused on the adoption and usage of electricity, skipping past the centuries-long effort to understand, capture, and transmit it. Electrification was not a single period of steady growth, nor a process through which any one product was adopted. Instead, it consisted of two separate waves of technological, industrial, and process-related transformation.

The first wave began around 1881, when Thomas Edison stood up electric power stations in Manhattan and London. Yet while Edison was quick to commercialize electricity—he had created the first working incandescent light bulb only two years earlier—demand for this resource was low. A quarter century after his first stations, an estimated 5% to 10% of mechanical drive power in the United States came from electricity (two-thirds of which was generated locally, rather than from a grid). But then, rather suddenly, the second wave began. Between 1910 and 1920, electricity's share of mechanical drive power quintupled to over 50% (nearly two-thirds of which came from independent electric utilities). By 1929, it stood at 78%.[4]

The difference between the first and second waves was not what portion of American industry used electricity, but the extent to which that portion did—and designed around it.[5]

When factories first adopted electrical power, it was typically used for lighting and to replace the on-premises source of power (usually steam). Owners did not rethink or replace the legacy infrastructure that would carry this power throughout the factory and put it to work. Instead, they continued to use a lumbering network of cogs and gears that were messy and loud and dangerous, difficult to upgrade or change, were either "all on" or "all off" (and therefore required the same amount of power to support a single operating station or the entire plant, and suffered from countless "single points of failure"), and struggled to support specialized work.

But eventually, new technologies and understandings gave owners both the reason and ability to redesign factories end-to-end for electricity, from replacing cogs with electric wires, to installing individual stations with bespoke and dedicated electrically powered motors for functions such as sewing, cutting, pressing, and welding.

The benefits were wide-ranging. The same factory now had considerably more space, more light, better air, and less life-threatening equipment. What's more, individual stations could be powered individually (which increased safety, while reducing costs and downtime), and they could use more specialized equipment, such as electric socket wrenches.

Factory owners could configure production areas around the logic of the production process, rather than hulking equipment, and even reconfigure these areas on a regular basis. These two changes meant that far more industries could deploy assembly lines (which had first emerged in the late 1700s), while those that already had such lines could extend them further and more efficiently. In 1913, Henry Ford created the first moving assembly line, which used electricity and conveyor belts to reduce the production time per car from 12.5 hours to 93 minutes, while also using less power. According to the historian David Nye, Ford's famous Highland Park plant was "built on the assumption that electrical light and power should be available everywhere."[6]

Once a few factories began this transformation, the entire market was forced to catch up, thereby spurring more investment and innovation in electricity-based infrastructure, equipment, and processes. Within a year of its first moving assembly line, Ford was producing more cars than the rest of the industry combined. By its 10 millionth car, it had built more than half of all cars on the road.

The "second wave" of industrial electricity adoption didn't depend on a single visionary making an evolutionary leap from Thomas Edison's core work. Nor was it driven merely by an increasing number of industrial power stations. Instead, it reflected a critical mass of interconnected innovations, spanning power management, manufacturing hardware, production theory, and more. Some of these innovations fit in the palm of a plant manager's hand, others needed a room, a few required a city, and all depended on people and processes. In aggregate, these innovations enabled what's known as the "Roaring Twenties," which saw the greatest average annual increases in labor and capital productivity in a hundred years, and propelled the Second Industrial Revolution.

## An iPhone 12 in 2008?

Electrification can help us better understand the rise of mobile. The iPhone *feels* like a starting point for the mobile era because it united or distilled all of the things we now think of as "the mobile internet"—touch screens, app stores, high-speed data, instant messaging—into a single product that we could touch, hold in the palm of our hands, and use throughout each and every day. But the mobile internet was created—and driven—by so much more.

It wasn't until the second iPhone, released in 2008, that the platform really began to take off, with sales increasing nearly 300% on a generational basis—a record that holds some 11 generations later. The second iPhone was the first to include 3G, which made the mobile web usable, and the App Store, which made wireless networks and smartphones useful.

Neither 3G nor the App Store were Apple-only innovations. The iPhone accessed 3G networks via chips made by Infineon that connected via standards led by groups such as the United Nations' International Telecommunication Union and the wireless industry's GSM Association. These standards were then deployed by wireless providers such as AT&T on top of wireless towers built by wireless tower companies such as Crown Castle and American Tower.

The iPhone had "an app for that" because millions of developers built them. These apps, in turn, were built on a wide variety of standards—from KDE to Java, HTML, and Unity—that were established and/or maintained by outside parties (some of whom competed with Apple in key areas). The App Store's payments worked because of digital payments systems and rails established by the major banks. The iPhone also depended on countless other technologies, from a Samsung CPU (licensed in turn from ARM), to an accelerometer from STMicroelectronics, Gorilla Glass from Corning, and other components from companies including Broadcom, Wolfson, and National Semiconductor. All of these creations and contributions, collectively, enabled the iPhone. They also shaped its improvement path.

We can see this in the iPhone 12, which was released in 2020 and was the company's first 5G device. Irrespective of Steve Jobs's brilliance, there was no amount of money that Apple could have spent to release the iPhone 12 in 2008. Even if Apple could have devised a 5G network chip back then, there were no 5G networks for it to use, nor 5G wireless standards through which to communicate to these networks, and no apps that took advantage of its low latency or bandwidth. Were Apple able to make its own ARM-like GPU back in 2008 (more than a decade before ARM itself), game developers (who generate 70% of

App Store revenues) would have lacked the game-engine technologies required to take advantage of its superpowered capabilities.

Getting to the iPhone 12 required ecosystem-wide innovation and investments, most of which sat outside Apple's purview, even though Apple's lucrative iOS platform was the core driver of these advancements. The business case for Verizon's 4G networks and American Tower Corporation's wireless tower buildouts depended on the consumer and business demand for faster and better wireless for apps such as Spotify, Netflix, and Snapchat. Without them, 4G's "killer app" would have been . . . slightly faster email. Better GPUs, meanwhile, were utilized by better games, and better cameras were made relevant by photo-sharing services such as Instagram. Better hardware powered greater engagement, which drove greater growth and profits for these companies, thereby driving better products, apps, and services.

In Chapter 9, I touched on the ways in which changing consumer habits, rather than just evolving technological capability, enables improvements in both hardware and software. A decade after the iPhone launched, Apple felt confident that it could remove the physical home button and instead ask device owners to return to the home screen and manage multitasking through touch-based swipes from the bottom of the screen. This new design opened up additional space inside the iPhone for more sophisticated sensors and computing components, and helped Apple (and its developers) introduce more complex, software based interaction models. Many video apps began to introduce gestures (for example, two fingers dragging up or down the screen) to increase or decrease volume, rather than require users to pause or litter the screen with unneeded buttons to do so.

## A Critical Mass of Working Pieces

With electrification and mobile in mind, we can confidently say only that the Metaverse will not suddenly arrive. There will be no clear "before Metaverse" and "after Metaverse"—only the ability to look back at a point in history when life was different. Some executives argue that we have already passed this threshold with the Metaverse. Their argument feels premature. Fewer than one in fourteen people today routinely engage with the virtual world—and these virtual worlds are almost exclusively games, have no meaningful interconnection (if any at all), with only marginal influence over society at large.

But *something* is happening. There is a reason why even the executives who think the Metaverse remains far off in the future, such as Zuckerberg, Sweeney, and Huang, believe now is the time to publicly commit to making it (a virtual) reality. As Sweeney has said, Epic Games has "had metaverse aspirations for a very, very long time. It started with text chat in realtime [*sic*] 3D with 300-polygon strangers. But only in recent years have a critical mass of working pieces started coming together rapidly."

These pieces include the proliferation of affordable mobile computers with high-resolution touch displays that are only a few inches away from two-thirds of everyone on earth over the age of 12. What's more, these devices are equipped with CPUs and GPUs capable of powering and rendering complex real-time rendered environments with dozens of concurrent users, each one steering their own avatar and capable of a wide range of actions. This functionality is furthered by 4G mobile chipsets and wireless networks that enable users to access these environments from wherever they are. The advent of programmable blockchains, meanwhile, has offered both the hope and mechanisms of harnessing the combined might and resources of every person and computer on earth to build not just the Metaverse, but a decentralized and healthy one.

Another piece is "cross-platform gaming," which has enabled users to play one another even if they use different operating systems (referred to as "cross-play"), buy virtual goods and currencies through any platform and then use them on another (cross-purchase), and carry their save data and in-game history across platforms (cross-progression).

These sorts of experiences have been technically possible for nearly two decades but were only enabled by the major gaming platforms (most notably, PlayStation) in 2018.

Cross-platform was essential in three ways. First, the very notion of a virtual persistent simulation that exists in the cloud is at odds with device-specific limitations. If the operating system you're using alters what you can see or do in "the Metaverse" and perhaps blocks you from visiting it altogether, there can be no "Metaverse" nor parallel plane of existence—instead, only software running on your device that lets you peer into one of several virtual realities. Second, the ability to use any device and interact with any other user led to a surge in engagement— just imagine how much less you might use Facebook if you had a different account with different friends and different photos on your PC versus on your iPhone, and if you could only message those who were using the same device as you. If the digital era has been defined by network effects and Metcalfe's Law, then the enablement of cross-platform play instantly made these virtual worlds more valuable by joining together their forked networks. Third, this increased engagement had a disproportionate impact on those building virtual worlds. Almost all of the costs to build a game, avatar, or item on *Roblox*, for example, are up-front and fixed. As a result, any increase in player spending dramatically increased an independent developer's profits, and thus their ability to reinvest in better or more games, avatars, and items.

We can also observe cultural changes. From its launch in 2017 through to the end of 2021, *Fortnite* generated an estimated $20 billion in revenue, the majority of which was from sales of digital avatars, backpacks, and dances (also known as "emotes"). *Fortnite* made Epic Games one of the largest sellers of fashion in the world, outgrossing giants such as Dolce & Gabbana, Prada, and Balenciaga by multiples, while also revealing that even "shooter" games were no longer just "games." The rise of NFTs throughout 2021, meanwhile, started to normalize the idea that purely virtual objects could be worth millions of dollars or more.

Relatedly, we should consider the ongoing destigmatization of time spent in virtual worlds, as well as the ways in which the COVID-19 pandemic accelerated this process. For decades, "gamers" have been making "fake" avatars and spending their free time in digital worlds while pursuing non-game-like objectives such as designing a room in *Second Life*, rather than killing a terrorist in *Counter-Strike*. A huge portion of society viewed such efforts as weird or wasteful or anti-social (if not worse). Some saw virtual worlds as the modern version of an adult man building a train set alone in his basement. Virtual weddings and funerals, which have been regular occurrences since the 1990s, were thought of as utterly absurd by most people—more of a punchline than something rather poignant.

It's difficult to imagine what could have more rapidly changed our perceptions of virtual worlds than time spent at home during the various COVID-19 lockdowns of 2020 and 2021. Millions of skeptics have now participated in (and enjoyed) virtual worlds and activities such as *Animal Crossing*, *Fortnite*, and *Roblox* as they sought out things to do, attended events once planned for the real world, or tried to spend time with their kids indoors. Not only have these experiences helped to destigmatize virtual life for society at large, they may even lead to another (older) generation participating in the Metaverse.*

The compounding impact of two years inside was profound. At the simplest level, the developers of virtual worlds benefited from more revenues, which in turn led to more investment and better products, thereby attracting more users and usage, thus more revenues, and so on. But as virtual worlds were destigmatized and it became clear that

---

* I see a number of similarities here to online groceries. Millions of consumers have known about online grocery service for years but refused to try it, even if they regularly bought clothes or toilet paper online. These holdouts simply believed that if someone else picked their groceries, they'd arrive spoiled, damaged, or in some indescribable way, just be "wrong." And there was no amount of marketing or endorsements to overcome this hesitancy. But the COVID-19 pandemic prompted many people to use grocery delivery for the first time, leading to the realization that online groceries are fine and the process is not just easy but nice. Some will go back to buying in person, but not all, nor all of the time.

everyone was a gamer, rather than just 13- to 34-year-old single men, the world's largest brands began to flock to the space and in doing so, further legitimize and diversify it. By the end of 2021, automotive giants (Ford), physical fitness brands (Nike), nonprofits (Reporters Sans Frontières), musicians (Justin Bieber), sports stars (Neymar Jr.), auction houses (Christie's), fashion houses (Louis Vuitton), and franchises (Marvel) had all made the Metaverse a key part of their business—if not the center of their growth strategy.

## The Next Drivers of Growth

What are the next "critical pieces" that might lead "Metaverse revenues" or "Metaverse adoption" to surge? One answer might be regulatory action against companies such as Apple and Google that forces them to unbundle their operating systems, software stores, payment solutions, and related services, and in doing so, compete individually in each area. Another popular answer is that we are waiting on an AR or VR headset that, like the iPhone, opens up the device category to hundreds of millions of consumers and many thousands of developers. Still more answers include blockchain-based decentralized computing, low-latency cloud computing, and the establishment of a common and widely adopted standard for 3D objects. Time will eventually reveal the truth, but for the foreseeable future, we can bet on three major drivers.

First, each of the underlying technologies required for the Metaverse is improving on an annual basis. Internet service becomes more widely available, faster, and less latent. Computing power, too, is more widely deployed, capable, and less costly. Game engines and integrated virtual world platforms are becoming easier to use, cheaper to build on, and more capable. The long process of standardization and interoperability is under way, driven in part by the success of integrated virtual world platforms and the crypto movement, but also by economic incentives. Payments, too, are slowly opening up through a mixture of regulatory action, lawsuits, and blockchains. Remember, Sweeney's "critical mass of working pieces" is not static, but constantly "coming together."

The second driver is the ongoing march of generational change. At the start of this book, I discussed the relevance of the "iPad-native" generation to the rise of *Roblox*. This group grew up expecting the world to be interactive—to be affected by their touch and their choices—and now that they can consume, the prior generations can see how different their behaviors and preferences are from that of older people. This is not new, of course. Depending on your own generational identity, you might have grown up sending postcards, spending hours each day after school talking on the phone, using instant messaging apps, or posting photographs on an online social network. The trajectory is clear. We know Generation Y games more than Gen X, Z more than Y, and Alpha more than Y. More than 75% of American children game on a single platform, *Roblox*. In other words, nearly everyone born today is a gamer. Which means 140 million new gamers are born globally each year.

The third driver is a result of how the first and second come together. Ultimately, the Metaverse will be ushered in through experiences. Smartphones, GPUs, and 4G didn't magically produce dynamic, real-time rendered virtual worlds—they needed developers and their imaginations. Note, too, that as the generation of "iPad-natives" ages, more people within it will shift from being consumers of or amateur hobbyists in virtual worlds to professional developers and business leaders in their own right.

## Chapter 13

# META-BUSINESSES

WHAT, THEN, MIGHT DEVELOPERS SOON PRODUCE? Throughout this book, I've avoided describing the "Metaverse in 2030" or offering any claims about what society will look like, overall, after the Metaverse arrives. The challenge with such broad prognostications is the feedback loops in between now and that date. An unforeseen technology will be created in 2023 or 2024 that in turn inspires new creations, or leads to new user behaviors, or manifests a new use case for that technology, leading to other innovations, changes, and applications, and so on. However, there are a few areas that will likely be transformed by the Metaverse in ways that in the short term, at least, can be said to be predictable. Millions if not billions of users and dollars will be drawn to the new experiences that result. With all the necessary caveats in mind, it is worth looking at what these transformations might look like.

## Education

The best example of impending transformation might be education. The sector is of critical importance to both society and the economy, and educational resources are scarce and starkly unequal in their distribution. It is also the leading example of what's known as "Baumol's Cost Disease," which refers to "the rise of salaries in jobs that have experienced no or low increase of labor productivity, in response to rising salaries in other jobs that have experienced higher labor productivity growth."[1]

This is not a critique of teachers. Rather, it reflects the fact that most jobs have become far more "productive," in economic terms, as a result of the many new digital technologies and developments over the past several decades. For example, an accountant has become far more efficient as a result of computerized databases and software such as Microsoft Office. An accountant today can do more "work" per unit of time, or manage more clients in the same amount of time, than an accountant could in the 1950s. The same is true for janitorial and security services, which now take advantage of more powerful motorized cleaning tools, or can monitor a facility using a network of digital cameras, sensors, and communications devices. Healthcare remains a labor-driven sector, but advances in diagnostics and therapeutic and life support technologies have helped to offset many of the costs associated with an aging population.

Teaching has seen a smaller increase in productivity compared to almost all other categories. A teacher in 2022 cannot, by most measures, teach more students than they could decades ago without adversely affecting the quality of their education. In addition, we have not found ways to teach for less time, either (that is, to teach faster). However, teaching salaries must compete with the salaries offered to someone who might otherwise become an accountant (or software engineer, or game designer), and must rise with the rising cost of living as a result of a growing economy. And beyond teacher time, education remains incredibly resource-intensive in terms of physical resources, from the size of the school, the quality of its facilities, and the quality of supplies. In fact, costs associated with these resources have partly increased due to new, more expensive technologies (for example, high-definition cameras and projectors, iPads, and so on).

The relative lack of productivity growth in education is demonstrated by its relative increases in costs. The US Bureau of Labor Statistics estimates that the cost of the average good in January 1980 has increased over 260% through January 2020, whereas the cost of college tuition and fees has grown 1,200%.[2] The second-closest sector, medical care and services, is up 600%.

While education has long lagged productivity growth in the West, technologists have been expecting it to beat most industry benchmarks. The assumption was that high school, colleges, and especially trade schools would be fundamentally reconfigured and displaced by remote learning. Many, if not most, students would learn remotely, not in the classroom but through on demand video, livestreamed classes, and AI-powered multiple choice. But among COVID's top lessons was that "Zoomschool" is terrible. There are many challenges when it comes to learning through a screen, but for the most part, we assume that we lose more than we might gain (or save financially).

The most obvious loss with remote learning is of "presence." When they are inside the classroom, students are in an education environment; they have agency and immersion that's totally unlike anything offered by a camera through which they can peer into an untouchable school set. Why presence matters is rather beside the point—but pedagogical research does show the clear benefits of sending students on field trips rather than limiting them to videos, of asking them to come to school rather than listen to recordings at home, and of encouraging them to learn "hands-on" whenever possible. The loss of presence entails the loss of everything from eye contact with (and scrutiny from) a teacher, the ability to co-learn alongside friends, and tactility, to the ability to build a hydraulic robot with syringes, use a Bunsen burner, and dissect a frog, fetal pig, or feral cat.

It is difficult to imagine at-home or at-distance education ever fully substituting for in-person education. But we are slowly closing the gap through new and predominantly Metaverse-focused technologies, such as volumetric display, VR and AR headsets, haptics, and eye-tracking cameras.

Not only are real-time rendered 3D technologies helping educators to take the classroom (and classmates) anywhere, but the rich virtual simulations that are on the horizon can greatly augment the learning process. At first, VR in the classroom was envisioned as little more than the ability to "visit" ancient Rome (incidentally, "visiting" Rome was long considered the "killer app" for VR headsets, but it turned out to be rather dull). Instead, students will "build Rome in a semester" and learn how aqueducts work by constructing them. Many students today and in past decades learned about gravity by watching their teacher drop a feather and a hammer, and then seeing a tape of Apollo 15 commander David Scott do the same on the moon (spoiler: they fall at the same speed). Such demonstrations need not go away, but they can be supplemented by the creation of elaborate, and virtual-only, Rube Goldberg machines, which students can then test under Earth-like gravity, on Mars, and even under sulfuric rainfalls of the Venetian upper atmosphere. Rather than create a volcanic eruption using vinegar and baking soda, students will immerse themselves in a volcano and then agitate its magma pools before they're both ejected into the sky.

Everything once imagined in *The Magic School Bus*, in other words, will become virtually possible—and at a greater scale, too. Unlike a physical classroom experience, these lessons will be available on demand, from anywhere around the world, and fully accessible (and more easily customized) to those students with physical or social disabilities. Some classes will include presentations from professional instructors whose live performances were motion captured and audio recorded. And as these experiences have no marginal costs—that is, they do not require extra time from a teacher, nor do they deplete supplies no matter how many times they're run—they can be priced at a fraction of the costs associated with learning that occurs in the classroom. Every student will be able to perform a dissection, no matter how wealthy their parents or the funding of their local school board. Indeed, these students will not even need to attend a school (and if they like, they'll be able to travel through the creature's various organ systems, rather than just cut them open).

Crucially, it will still be possible for these virtual classes to be supplemented by a dedicated, live teacher. Imagine the "real" Jane Goodall reproduced in a virtual environment and guiding students through

Tanzania's Gombe Stream National Park, with these students' "home-room" teacher joining in and further personalizing the experience. The costs involved with such an experience will be a fraction of that involved in a real field trip—certainly one to Tanzania—and may even offer more than such a trip could.

None of this is to suggest that education involving VR and virtual worlds will be easy. Pedagogy is an art, and learning is hard to measure. But it's not difficult to imagine how virtual experiences might enhance learning while also expanding access and reducing its costs. There will be less of a gap between in-person and at-distance educations, competitive marketplaces for pre-made lessons and live tutors, and exponentially greater reach for great teachers and their work.

Careful readers will note that such experiences do not by themselves make, nor require, the Metaverse. It's possible for compelling real-time rendered 3D worlds focused on education to exist without the Metaverse. However, interoperation between these experiences and all others, as well as the real world, is of obvious value. If users can bring their avatars to these worlds, they're likely to use them more often. If their educational account history can be written "in school," and then read and expanded upon elsewhere, learners will be more likely to keep learning and their experiences will be more richly personalized.

## Lifestyle Businesses

Education is just one of many socially focused experiences that will be transformed by the Metaverse. Today, millions of people exercise each day using digital services such as Peloton, which offers live and on demand video-based cycling classes with gamified leaderboards and high-score tracking, and Mirror, a Lululemon subsidiary that boasts a wider range of fitness routines delivered by a partly transparent instructor projected through a reflective mirror. Peloton has since expanded into real-time rendered virtual games, such as *Lanebreak*, where a cyclist controls a wheel rolling across a fantastical track to earn points and dodge obstacles. This is a sign of things to come; perhaps sometime soon, our morning routine will involve our *Roblox* avatar cycling across the snowy *Star Wars* planet of Hoth through a Peloton application on our Facebook VR headset, all while chatting with our friends.

Mindfulness, meditation, physiotherapy, and psychotherapy are likely to be similarly altered, by a mix of electromyographic sensors, volumetric holographic displays, immersive headsets, and projection and tracking cameras that collectively provide support, stimulation, and simulation never before possible.

Dating is another fascinating category when considering the impact of the Metaverse. Prior to the launch of Tinder, some believed that online dating had been "solved"—all one had to do was fill out dozens to hundreds of multiple-choice quizzes that would be crunched into a mystery compatibility score through which two would-be lovebirds would be matched. But this belief and the companies built on it were disrupted by a photo-based model in which users "swipe right" or "swipe left" to see if there's a shared interest in chatting, and with the average user spending between three to seven seconds making such a choice.[3] In recent years, dating applications have added new features for matched couples, such as casual games and quizzes, voice notes, and the ability to share their favorite playlists on Spotify and Apple Music. In the future, dating applications will likely offer couples a variety of immersive virtual worlds which help a would-be pairing get to know one another. These might span simulated reality ("dinner in Paris") or the fantastical ("dinner in Paris . . . on the Moon"), include live performances from motion captured avatars* (imagine mariachis or attending a digital twin of London's Royal Ballet, but from Atlanta), and potentially lead to reinventions of classic game-show formats such

---

* Neal Stephenson described this sort of technology and experience at length in *The Diamond Age*, which was published in 1995, three years after *Snow Crash*. He called such products interactive books, or "ractives" for short, with performers known as "ractors," as in interactive actors.

as *The Dating Game*. It's also likely these apps integrate into third-party virtual worlds (this is the Metaverse, after all), enabling, as an example, a matched couple to easily jump into a virtual Peloton or Headspace-based experience.

## Entertainment

It's increasingly common to hear that the future of "linear media" such as films and TV shows is VR and AR. Rather than watch *Game of Thrones* or the Golden State Warriors play the Cleveland Cavaliers on our couch sitting in front of our 30 × 60–inch flatscreen, we will put on a VR headset and watch shows on simulated IMAX-sized screens, or sit courtside—with our friends sitting beside us. Alternatively, we might watch via augmented reality glasses that make it seem like we still have a living room TV. The films and TV shows, of course, will be filmed for 360° immersion. When Travis Bickle says "You talkin' to me?," you can be virtually standing in front of, or even behind, him.

These predictions remind me of how many once envisioned newspapers like the *New York Times* would be altered by the internet.[4] In the 1990s, some believed that "in the future" the *Times* would send a PDF of each day's edition to every subscriber's printer, which would then dutifully print it before its owner woke up—thereby obviating the need for costly printing presses and elaborate home delivery systems. The more daring theorists imagined this PDF might even exclude sections the individual reader did not want, thereby saving both paper and ink. Decades later, the *Times* does offer this option, but almost no one uses it. Instead, subscribers access a constantly changing and never-printed online copy of the paper that has no clear divisions between sections and essentially cannot be read "front to back." Most news readers don't even start with a newspaper at all. Instead, they consume their news via aggregator solutions such as Apple News, and social media newsfeeds which intermingle countless stories from disparate publishers, alongside photos of your friends and family.

The future of entertainment will probably involve similar remixing. "Film" and "TV" will not go away—just as oral storytelling, serials, novels, and radio shows still exist centuries after they were first created—but we can expect rich interconnection between film and interactive experiences (broadly considered "games"). Facilitating this transformation is the increasing use of real-time rendering engines, such as Unreal and Unity, in filmmaking.

Historically, movies such as *Harry Potter* or *Star Wars* have used non-real-time rendering software. There was no need to produce a frame in milliseconds during the production process and so it made sense to spend more time (anywhere from one additional millisecond to several days) making the image look more realistic or detailed. In addition, the goal of the computer graphics department was to virtually produce an already-known image (that is, one based on a storyboard). As such, moviemakers didn't need to "build Manhattan" or even a single street in the West Village in order to support a set piece in *The Avengers*, least of all a street that could simulate the "real New York" and anything that might happen to it when aliens invade and Infinity Stones are involved.

But over the past five years, Hollywood has progressively integrated real-time rendering engines, most typically Unity and Unreal, into their filming process. For 2019's *The Lion King*, a purely CGI-based film but one that was designed to look like "live action," the director Jon Favreau immersed himself into each scene through a Unity-based re-creation, often while wearing a VR headset. This allowed him to understand a purely virtual set as though it were a typical "real world" film shoot—a process that he claims aided everything from where to place and angle a shot, to how the camera would track its fictional leads, as well as the lighting and coloring of the environment. The final rendering was still produced in Maya, non-real-time animation software published by Autodesk.

Building upon his work on *The Lion King*, Favreau helped pioneer "virtual production" stages where an enormous circular room is constructed using walls and ceilings made of high-density LEDs (the rooms themselves are called "volumes"). The LEDs were then lit up with Unreal-based real-time renders. This innovation provided a number of benefits. The simplest was that it allowed everyone inside the volume to experience what Favreau did in VR, but without wearing a headset. It also meant that "real people" could be seen inside the environment too—rather than everyone just watching pre-planned animations of *Timon & Pumbaa*. In addition, the cast could be affected by the volume's LEDs; the light shining down from a virtual sun would recolor an actor directly and provide them with an accurate shadow—it wouldn't need to be applied or corrected in "post-production." A set could have the perfect sunset year-round—and years later that exact same setting could be reproduced in seconds.

One of the leaders in virtual production is Industrial Light & Magic, the visual effects company founded by *Star Wars* creator George Lucas and that is now owned by Disney. ILM estimates that when a film or series is designed for LED volumes, it's possible to film 30% to 50% faster than when shooting through a mixture of "real world" and "green screen" sets, and that postproduction costs are lower, too. ILM points to the hit *Star Wars* TV series, *The Mandalorian*, which was created and directed by Favreau and cost roughly one quarter as much per minute as the typical *Star Wars* film (it was also better received by both critics and viewers). Nearly all of the show's first season—which spanned an unnamed ice world, the desert planet Nevarro, the forested Sorgan, deep space, and dozens of subsets in each—was shot on a single virtual stage in Manhattan Beach, California.

What does virtual production have to do with the Metaverse beyond the use of similar engines and virtual worlds? The connections start with "virtual backlots." If you visit Disney's physical studio backlot, you'll find stages and lockers full of old *Captain America* costumes, miniature models of the Death Star, and the literal living rooms of *Modern Family*, *New Girl*, and *How I Met Your Mother*. Now, Disney's servers are being filled with virtual versions of every 3D object, texture, outfit, environment, building, facial scan, and anything else it has made. This doesn't just make it easier to film a sequel—it makes it easier to make all derivative works. If Peloton wants to sell a course set on the Death

Star or the Avengers' Campus, it can repurpose (in other words, license) much of what Disney has made. If Tinder wants to offer virtual dates on Mustafar, the same applies. Instead of playing blackjack via the video-based iCasino, why not play on Canto Bight? Rather than launch a *Star Wars* integration in *Fortnite*, Disney will just populate their own mini-worlds on *Fortnite Creative* using what they've already built.

These won't just be opportunities to personally experience the filmed world of *Star Wars*, either. They will become a core part of the storytelling experience. In between weekly episodes of *The Mandalorian* or *Batman*, fans will be able to join their heroes in canonical (or noncanonical) events and side missions. At 9 p.m. on a Wednesday night, for example, Marvel might tweet that the Avengers "need our help," with Tony Stark, as live-performed by Robert Downey Jr. (or perhaps someone who bears little resemblance to him but steers an avatar that does), leading the way. Alternatively, fans will have the opportunity to live out what they watched in a movie or show. The end of *The Avengers: Age of Ultron* in 2015 involved the titular heroes fighting a legion of evil robots on a chunk of land floating above the earth. In 2030, players will have the chance to do the same.

Similar opportunities will open up to sports fans. We may use VR to sit virtually courtside, but it's more likely the games we watch are nearly instantaneously captured and reproduced into a "video game." If you own NBA 2K27, you'll be able to jump into a specific moment from a game that finished only minutes earlier and then see if you could have won the game—or at least made the shot that a star player didn't. Sports fandom is currently isolated between watching a game, playing a sports video game, participating in fantasy sports, making online wagers, and buying NFTs, but we'll likely find that each of these experiences melds together and in doing so, creates new ones.

Betting and gambling will be transformed as well. There are already tens of millions of people placing online wagers, using Zoom-based casinos, or enjoying game-based casinos such as the Be Lucky: Los Santos in *Grand Theft Auto*. In the future, many of us will go to Metaverse casinos where we're served by live, motion capture–powered dealers while enjoying live, motion capture–powered musical performances. Or recall *Zed Run* from Chapter 11. Each week, hundreds of thousands of dollars are bet on its virtual horse races, with many of these horses worth millions. The economy of *Zed Run* is upheld through its blockchain-based programming, which provides bettors with the trust that the races are not rigged and horse owners with the faith that the "genes" of their virtual horses will be programmatically passed on when they're bred.

Others are reimagining entertainment at a more abstract level. From December 2020 to March 2021, Genvid Technologies hosted a "Massively Interactive Live Event" (MILE) on Facebook Watch, called *Rival Peak*. The title was a sort of virtual mashup of *American Idol*, *Big Brother*, and *Lost*. Thirteen AI contestants were trapped in a remote part of the Pacific Northwest, and the audience could watch them interact, fight to survive, and uncover various mysteries through dozens of cameras running 24 hours a day for all 13 weeks. While the audience could not directly control a given character, they could still affect the simulation in real time—solving puzzles to aid a given hero or create an obstacle for a villain, weighing in on the choices of the AI characters, and voting on who would be booted off the island. Though visually and creatively primitive, *Rival Peak* is an indication of what the future of live interactive entertainment could look like— that is, not supporting linear stories, but collectively producing an interactive one. In 2022, Genvid launched *The Walking Dead: The Last M.I.L.E.* with the comic book franchises, Robert Kirkman, and his company, Skybound Entertainment. The experience allows viewers, for the first time, to decide who lives and who dies in *The Walking Dead*, while also steering competing factions of humans toward, or away from, conflict. Audience members can also design their own avatars, who will then be released into the world and folded into the story. What might come next? Well, most of us don't want a real "Hunger Games," but it might be fun to watch a high fidelity real-time rendered version played by our favorite actors, sports stars, and even politicians, each of whom participates via avatar.

## Sex and Sex Work

Changes to the sex work industry are likely to be even more profound than those experienced by Hollywood, and in the process, further blur the line between pornography and prostitution. In 2022, one can hire a sex worker for a private online show and even take control of their smart sex toys (or provide them with control over yours). What might this look and feel like with an ever-growing number of internet-connected haptic devices, improvements in real-time rendering, immersive AR and VR headsets, and high-concurrency GPUs? Some of the results are relatively easy to imagine ("Sex, but in VR!"), others less so. Recall from Chapter 9 how armbands from CTRL-labs could use electromyography to reproduce precise finger movements—or to map the muscle movements used to move a finger to an entirely different motion, such as controlling the legs of a spider. With that in mind, what is sex experienced through an ultrasonic force field? Or when five, 100, or 10,000 "concurrent users" combine to construct some form of real-time rendered, mixed reality orgy, rather than a concert or battle royale?

Of course, such experiences raise the potential for considerable abuse (more on this soon), but also questions of platform power. None of the major mobile or console computing platforms enable for sex or pornography-based applications. PornHub.com, which typically ranks among the 70–80 most used websites in the world; Chaturbate, which ranks in the top 50; and OnlyFans, which ranks in the top 500 but whose revenue exceeds that of The Match Group (owners of Tinder, Match.com, Hinge, PlentyofFish, OkCupid, and more), are not permitted in the iOS or Android app stores. The justification for prohibition varies. Steve Jobs once told a user that Apple does "believe we have a moral responsibility to keep porn off the iPhone," though some speculate these policies are intended to avoid liability and the optics of taking a commission from sex work. The result doubtlessly harms individual sex workers—as I've mentioned often throughout this book, applications strongly outperform browser-based experiences in terms of usage and monetization—though pornography, as a category, still thrives. Videos and photos work well enough from a mobile web browser, and by and large, consumers are not deterred by the need to use them.

But as we've seen, richly rendered VR and AR experiences are essentially impossible via mobile web browsers. Accordingly, the policies of Apple, Amazon, Google, PlayStation, and others effectively block the entire category's advancement. Some might see this as a good thing; others could argue it deprives sex workers of higher incomes and greater safety.

## Fashion and Advertising

For the past 60 years, virtual worlds have been largely ignored by advertisers and fashion houses. Today, less than 5% of video gaming revenue comes from advertising. In contrast, most major media categories, such as TV, audio (inclusive of music, talk radio, podcasts, and so on), and news generate 50% or more of their revenues from advertisers, rather than audiences. And although hundreds of millions entertain themselves in the virtual worlds each year, 2021 was the first time that brands such as Adidas, Moncler, Balenciaga, Gucci, and Prada saw these spaces as deserving of any real attention. This will need to change.

Advertising in virtual spaces is difficult for a few reasons. First, the gaming industry was "offline" for the first several decades and each title took years to produce. As a result, there was no way to update a game's in-game advertising, meaning any placed ads could quickly become out of date. This is also why books typically lack ads, save for those promoting the author's other works, even though newspapers and magazines historically relied upon them. Ford won't pay much for an ad that, for most readers, is touting the "specs" of an old car (Ford would probably consider such impressions harmful). Technical limitations of this kind no longer exist for video games, as they can now be updated over the internet, but the cultural consequences endure. With the exception of casual mobile games like *Candy Crush*, the gaming community is largely unfamiliar with and highly resistant to in-game advertising. Even though few consumers of television, print magazines and newspapers, and radio enjoy the ads that often litter these mediums, ads have always been part of the expereince.

The bigger issue might be determining what an ad is or should be in a real-time rendered 3D virtual world—and how to price it and sell it. For much of the 20th century, most ads were individually negotiated and placed. That is, someone at a company like Procter & Gamble would work with someone at CBS so that an Ivory Soap ad would air as the first commercial in the second ad block in the 9 p.m. airing of *I Love Lucy* and at a specific price. Most digital advertising today is done programmatically. For example, advertisers will say who they want to target, with what ads (a banner image, a sponsored social media post, a sponsored search result, and so on), up until a certain amount of money has been spent at a given cost-per-click or a set amount of time has transpired.

Finding the core "ad unit" for 3D-rendered virtual worlds is a challenge. Many games have in-game billboards, including the PlayStation 4 game *Marvel's Spider-Man*, which is set in Manhattan, and the cross-platform hit *Fortnite*. However, their implementations are quite different. The size of these posters might vary by multiples, meaning a different image would likely be needed for one versus another (whereas Google Ad Words work regardless of screen size). In addition, players might pass by these posters at varying speeds, from varying distances, and in various situations (a leisurely walk versus an intense firefight). All of this makes it hard to value either game's billboards, let alone buy them programmatically. There are many other potential ad units inside a virtual world—commercials played by in-game car radios, virtual soft drinks branded like real-world ones—but these are even harder to design for and measure. Then there are the technical complexities of inserting personalized ads into synchronous experiences, determining when an ad should be shared with your friends or not (it makes sense for the whole squad to see a banner for the next Avengers movie, but not necessarily for a medicinal cream), and so on.

Augmented reality advertising is conceptually easier, as the canvas for said ads is the real world rather than myriad virtual ones, but the execution is perhaps even harder. If users are inundated with unprompted or obtrusive ads overlayed atop the real world, they'll change headsets. The risk of these ads causing an accident is also high.

In the United States, advertising expenditures have comprised 0.9% to 1.1% of GDP for more than a century (with temporary exceptions during the world wars). If the Metaverse is to be a major economic force, ad buyers will have to find a way to be relevant in it and the ad tech industry will eventually figure out how to offer and adequately measure programmatic ads placed across myriad virtual spaces and objects in the Metaverse.

Still, some argue that the Metaverse will require a more fundamental rethinking of how to advertise a given product.

In 2019, Nike built an immersive *Fortnite Creative Mode* world under the Air Jordan brand, entitled "Downtown Drop." In it, players raced through the streets of a fantastical city while wearing rocket-powered shoes, performing tricks and collecting coins to beat other players. While players could purchase and unlock exclusive Air Jordan avatars and items during and through this "limited time mode," the goal of "Downtown Drop" was to express the ethos of Nike's Air Jordan— for players to know what the brand felt like, no matter the medium. In September 2021, Tim Sweeney told the *Washington Post* that a "carmaker who wants to make a presence in the metaverse isn't going to run ads. They're going to drop their car into the [virtual] world in real time and you'll be able to drive it around. And they're going to work with lots of content creators with different experiences to ensure their car is playable here and there, and that it's receiving the attention it deserves."[5]

Needless to say, dropping a new, drivable car model into a virtual world is much trickier than placing market copy into targeted search results, telling a compelling 30-second or two-minute story in a commercial, or producing a "native advertisement" with a YouTuber. It requires building experiences and virtual products that users actively choose to engage with and use in lieu of the entertainment they originally sought out. And almost no ad agencies or marketing departments today have even the basic skillsets required to build such experiences. Still, the likely profits from successful advertising in the Metaverse, the necessity of differentiation, and the lessons of the consumer internet era seem likely to inspire significant experimentation in the years to come.

Upstart brands such as Casper, Quip, Ro, Warby Parker, Allbirds, and Dollar Shave Club didn't just take advantage of direct-to-consumer e-commerce models—they also won market share from long-standing incumbents through novel marketing techniques such as search engine optimization, A/B testing, and referral codes, and developing unique social media identities. But in 2022, these strategies are not novel— they're commodity, table-stakes, dull. They enable no brand, new or old, to find new audiences or stand out. Virtual worlds, however, remain largely unconquered territory.

For the same reasons, today's fashion brands will also need to "enter the Metaverse." As more of human culture shifts into virtual worlds, individuals will seek out new ways to express their identities and show off. This is demonstrated clearly through *Fortnite*, which has spent several years generating more revenue than any other game in history, and primarily monetizes through the sale of cosmetic items (and as I mentioned earlier, these revenues exceed many of the top fashion labels, too). NFTs reiterate this as well. The most successful NFT collections are not for virtual goods nor trading cards but identity- and community-oriented "profile pictures" such as Cryptopunks and Bored Apes.

If today's labels do not meet this need, new labels will emerge which will replace them. In addition, the Metaverse will place pressure on the physical sales of many companies, such as Louis Vuitton and Balenciaga. If more work and leisure occur in virtual spaces, then we'll need fewer purses and probably spend less on those we do buy. But to this end, these labels will likely use their physical sales to facilitate and bolster the value of their digital ones. For example, a consumer who buys a physical Brooklyn Nets jersey or Prada bag might also get the rights to a virtual or NFT simulacra, or a discount when buying one. Or perhaps only those who do buy "the right thing" can get a digital copy. In other cases, a digital purchase might lead to a physical one. Our identities, after all, are not purely online or offline, physical or metaphysical. They persist, just like the Metaverse.

# Industry

In Chapter 4, I highlighted how and why the Metaverse would start with consumer leisure and then move into industry and enterprise, rather than the reverse, as happened with prior computing and networking waves. The expansion into industry will be slow. The technical requirements for simulation fidelity and flexibility are much higher than in games or film, while success ultimately depends on reeducating employees who have been trained around now-legacy software solutions and business processes. And to start, most "Metaverse investments" will be premised upon hypotheses, rather than best practices—meaning investments will be constrained and the profits often disappointing. But eventually, and with the current internet, much of the Metaverse and its revenues will exist and occur out of sight from the average consumer.

Consider, as an example, the 56-acre, 20-building, multi-billion-dollar redevelopment of Water Street in Tampa, Florida. As part of this project, Strategic Development Partners produced a 17-foot-diameter, 3D-printed, and modular scale model of the city, which was then supported by twelve 5K laser cameras that projected 25 million pixels atop this model, based on city data feeds for weather, traffic, population density, and more. All of this was run by an Unreal-based real-time rendered simulation that could be viewed through a touchscreen or VR headset.

The perks of such a simulation are difficult to describe in writing for the very reason SDP saw value in building a physical model and 3D digital twin in the first place. However, SDP enabled the city, prospective tenants, and investors, as well as construction partners, to understand and plan for the project in unique ways. It was possible to see exactly how present-day Tampa would be affected by the construction process, as well as by the completed project. How would a five-year build affect local traffic and how would the effects be different from those of a six-year build? What would happen if a given building were replaced by a park, or its floors reduced from 15 to 11? How would the views of other buildings and parks in the area be affected by the development, including through refracted light or radiated heat—and at any time or day in the year? How would these buildings shape emergency response times in the area? Might they require a new police, fire, or ambulance station? On which sides of the buildings should a fire escape be built?

Today, these simulations are primarily used to design and understand a building or project. Eventually, they will be used to operate the resulting buildings and the businesses they house. For example, the signage (physical, digital, and virtual) inside a Starbucks will be selected and altered based on real-time tracking of which sorts of customers use the store and when, as well as remaining inventory in that location. The mall where a Starbucks is located will also direct customers to that site, or discourage them from doing so, based on its lines and the proximity of substitutes (or another Starbucks). And the mall will connect into the city's underlying infrastructure systems, thereby enabling AI-powered traffic light networks to operate with more (that is, better) information, and helping city services, such as fire and police, better respond to emergencies.

Though these examples focus on what's called "AEC," or architecture, engineering, and construction, such ideas are easily repurposed to other use cases. Various militaries around the world have been using 3D simulations for years—and as discussed in the hardware chapter, the US Army awarded Microsoft a contract worth more than $20 billion for HoloLens headsets and software. The utility of digital twins in aerospace and defense companies, too, is obvious (if perhaps even more terrifying than the army using VR). More hopeful is medicine and healthcare. Just as students might use 3D simulation to explore the human body, so too will physicians. In 2021, neurosurgeons at Johns Hopkins performed the hospital's first-ever AR-surgery on a live patient. According to Dr. Timothy Witham, who led the surgery and is the director of the hospital's Spinal Fusion Laboratory, "It's like having a GPS navigator in front of your eyes in a natural way so you don't have to look at a separate screen to see your patient's CT scan."[6]

Dr. Witham's GPS analogy reveals the critical difference between the so-called minimum viable product of commercial AR/VR and that for consumer leisure. To gain adoption, consumer VR/AR headsets must be more compelling or functional than the experiences offered by alternatives, such as a console video game or smartphone messaging app. The immersion offered by mixed-reality devices is a differentiator, but as discussed in Chapter 9, there are still many drawbacks. For example, *Fortnite* can be played on nearly any device, which means a user can play with anyone they know. *Population: One* is essentially limited to those who own a VR headset. In addition, *Fortnite* can also be experienced at a higher resolution, with greater visual fidelity, higher frame rates, more concurrent users, and without the risk of nausea. For many gamers, VR games are not yet good enough to successfully compete with console-, PC-, or smartphone-based titles. But comparing surgery with AR to surgery without it is like comparing driving with GPS to driving without it—the trip will be made regardless of whether the technology exists, while its use depends on whether it has a meaningful impact on the outcome (e.g., a shorter drive time). For surgery, this means a higher success rate, faster recovery time, or lower cost. And while the technical limitations of today's AR/VR devices doubtlessly limit their contributions to surgery, even a slight impact will justify their cost and use.

*Chapter 14*

# METAVERSE WINNERS AND LOSERS

IF THE METAVERSE IS A "QUASI-SUCCESSOR STATE" to the mobile and cloud era of computing and networking, and will eventually transform most industries and reach nearly every person on earth, a few very broad questions must be addressed. What will the value of a new "Metaverse economy" be? Who will lead it? And what will the Metaverse mean for society?

## The Economic Value of the Metaverse

Although corporate executives can't yet agree on exactly what the Metaverse is and when it will arrive, most believe it will be worth multiple trillions of dollars. Nvidia's Jensen Huang predicts the value of the Metaverse will eventually "exceed that" of the physical world.

Trying to project the size of the Metaverse economy is a fun, albeit frustrating, exercise. Even by the time the Metaverse is "here," there will likely be no consensus on its value. After all, we are at least 15 years into the mobile internet era, nearly 40 years into the internet era, and more than three-quarters of a century into the digital computing era, yet have no consensus answer for how much the "mobile economy," "internet economy," or "digital economy" might be worth. In fact, it's rare that anyone even tries to value any of them.* Instead, most analysts and journalists just sum the valuations or revenues of the companies which primarily support these loosely defined categories. The challenge in trying to measure any of these economies is that they aren't really an "economy." Instead, they are collections of technologies which are deeply intertwined with and dependent upon the "traditional economy," and as such, trying to value their would-be economy is more of an art of allocation, rather than a science of measurement or observation.

Consider the book you're reading now. Odds are that you purchased it online. Does the money you paid for it count as "digital revenue," even though it was physically produced, physically distributed, and is being physically consumed? Should *some* of your purchase be digital, and if so, how much and why? How does the ratio change if you're reading an e-book? What if you were boarding a plane, realized you wouldn't have anything to do while on the flight, and used your iPhone to download a digital audio–only copy—does that change the split? What if you only knew about the book from a Facebook post? Does it matter if I wrote the book using a cloud-based word processor rather than an offline one (or, dare I say, by hand)?

Things get even more difficult when we think of subsets of digital revenue, such as internet revenues or mobile revenues, both of which are likely to be the closest methodological comparison to "the Metaverse economy." Does Netflix, an internet-based video service, have mobile revenues? The company does have some mobile-only subscribers, but isolating the revenue of these customers as "mobile revenues" does not address the revenues of subscribers who use mobile devices to watch Netflix some, but not all, of the time, and pay to access the service across all sorts of devices. Should "mobile" be allocated a share of a monthly subscription fee based on its share of a user's time? Doesn't that mean that a user places equivalent value on watching a film on a 65" living room TV screen as they do a 5" × 5" smartphone used on the subway? Is a Wi-Fi-only iPad that never

---

* In case such efforts do seem familiar, it's probably because I've mentioned several estimates throughout this book.

leaves the home a "mobile" device? Probably, but why isn't a smart television that connects to Wi-Fi considered a mobile device? And can you even say there are "mobile" broadband revenues when the bits they transmit primarily travel through fixed-line cables? For that matter, isn't it true that most "digital devices" purchased today would not have been bought were it not for the internet? When Tesla updates a car's software over the internet in order to improve battery life and/or charging efficiency, how, exactly, should this value be counted or measured?

We can see some presages of these issues now. If you upgrade from a three-year-old iPad to a newer iPad Pro solely for its GPU in order to engage in high concurrent user real-time rendered 3D virtual worlds, what is the Metaverse allocation? If Nike sells sneakers with a bundled NFT or *Fortnite* edition, are there Metaverse revenues, and if so, how much? Is there an interoperability threshold for virtual goods to be considered Metaverse purchases, rather than just video game items? If you bet in US dollars on a blockchain horse, or cryptocurrency on a real one, is there a difference? If, as Bill Gates imagines, most video calls on Microsoft Teams shift to real-time rendered 3D environments, what portion of its subscription fee falls under "Metaverse"? If a building is operated through a digial twin, what part of its expenses should be counted? When broadband infrastructure is replaced by higher-capacity, real-time delivery, is that "Metaverse investment"? Nearly all the applications that will use and benefit from this leap have little to do with the Metaverse, at least today. Yet the drivers of the investment in low-latency networking are the few experiences that require it: synchronous real-time rendered virtual worlds, augmented reality, and cloud game streaming.

While the questions described above are helpful thought exercises, they have no single answer. It's particularly challenging to weigh in on those focused on the Metaverse, which doesn't yet exist and won't have an obvious start date. With this in mind, the more practical approach to sizing the "Metaverse economy" is to be more philosophical.

For nearly eight decades, the digital economy's share of the world economy has grown. The few estimates that do exist suggest roughly 20% of the world economy is now digital, which would value the latter at roughly $19 trillion in 2021. In the 1990s and early 2000s, most but not all of the digital economy's growth was driven by the proliferation of PCs and internet service, while the following two decades were primarily but not exclusively from mobile and cloud. These latter two waves meant that digital businesses, content, and services could be accessed by more people, in more places, more often and more easily, while supporting new use cases. The mobile and cloud waves also came to eclipse everything that preceded them. In most cases, "digital revenues" are not new. The dating services industry, for example, was negligible in size before the internet, and then grew by orders of magnitude through mobile. The recorded music industry more than doubled through digital compact discs, but then fell 75% through internet-based delivery.

The arc of the Metaverse will be broadly similar. Overall, it will help grow the global economy, even as it shrinks parts of it (commercial real estate, perhaps). In doing so, digital's share of the global economy will increase, as will the Metaverse's share of digital's share.

Granting this assumption allows us to do some modeling. If the Metaverse is, say, 10% of digital by 2032, and digital's share of the world economy grows from 20% to 25% over that time, and the world economy continues to grow at an average of 2.5%, then in a decade, the Metaverse economy would be worth $3.65 trillion annually. This figure would also indicate that the Metaverse constituted a quarter of the growth in the digital economy since 2022, and nearly 10% of real GDP growth over that same time (much of the rest would stem from population increases and shifting consumer habits, such as buying more cars, consuming more water, and so on). At 15% of the digital economy, the Metaverse would be $5.45 trillion annually, a third of digital's growth, and 13% of the world economy's growth. At 20%, it would be $7.25 trillion, half, and one-sixth. Some imagine the Metaverse might be as much as 30% of the digital economy of 2032.

However speculative, the above exercise describes exactly how the economy is transformed. Those who pioneer in the Metaverse will be over-indexed to the young, grow faster than the companies leading in either the "digital" or "physical" economy, and redefine our business models, behaviors, and culture. In turn, venture and public-market investors will more highly value these companies than the rest of the market, thereby producing many trillions more in wealth to those who create, work at, or invest in these companies.

A precious few of these companies will become critical intermediaries between consumers, businesses, and governments—multi-trillion-dollar companies in their own right. That's the odd thing about saying the digital economy is 20% of the world economy. No matter how sound the methodology, the conclusion skips over the fact that most of the remaining 80% is digitally powered or informed. This is also why we recognize the big five technology giants as being even more powerful than their revenues alone suggest. Google, Apple, Facebook, Amazon, and Microsoft combined reported revenues of $1.4 trillion in 2021, less than 10% of total digital spending, and 1.6% of the total world economy. However, these companies have a disproportionate impact on all of the revenues they don't recognize on their balance sheet, take a cut of many of them (for example, via Amazon's data centers or Google's ads), and sometimes set their technical standards and business models, too.

## How Today's Tech Giants Are Positioned for the Metaverse

Which companies will lead in the Metaverse era? History can inform how we answer this question.

There are five categories through which we can understand corporate trajectories. First, countless new companies, products, and services will be developed, ultimately affecting, reaching, or transforming nearly every country, consumer, and industry. Some of the new entrants will displace today's leaders, which will either perish or decline into irrelevancy. Examples here include AOL, ICQ, Yahoo, Palm, and Blockbuster (the second category). Some displaced giants actually expand as a result of the overall growth in the digital economy. IBM and Microsoft have never had a smaller market share of computers, yet each is more valuable than at any point during their supposed heydays. A fifth category of companies will ward off displacement and disruption, and grow their core businesses. So who might be the case studies of the shift to the Metaverse?

Facebook, unlike MySpace, successfully navigated the transition to mobile. But the company must transform again, and at a time when regulators seem unlikely to support acquisitions similar to those of Instagram and WhatsApp, which facilitated the company's pivot to mobile, and Oculus VR and CTRL-labs, which laid the foundation for its Metaverse plans. The company also faces strategic blocks from the hardware-based platforms upon which its services typically run— and, at the same time, its reputation has never been so negative. Still, it would be a mistake to discount Facebook. The social networking giant has three billion monthly users, two billion daily users, and the most used identity system online. It already spends $12 billion per year on Metaverse-related initiatives (and generates over $50 billion per year in cash flow on close to $100 billion in revenue), has a multi-year head start shipping VR hardware, and an in-control founder who believes in the Metaverse as much as any corporate executive.

But just as one can't count out Facebook, investment and conviction do not alone ensure success. Disruption is not a linear process, but a recursive and unpredictable one. And as we've seen, there is a lot of confusion and open questions surrounding the Metaverse. When will key technological advances arrive? How are they best realized? What's the ideal monetization model for it? What new use cases and behaviors will be created as a result of new technology? In the 1990s, Microsoft believed in both mobile and the internet and had many of the products, technologies, and resources needed to build what Google, Apple, Facebook, and Amazon did instead. Microsoft, it turned out, was wrong about everything from the role of app stores and smartphones to the importance of touchscreens to everyday consumers, and was distracted by the need to maintain its hugely successful Windows operating system and integrated Microsoft Exchange, Server, and Office suites. The Microsoft that is so valuable today is a result of a decision to finally shed its attachments to its own stacks and suites and instead support what the customer preferred.

In many categories, Microsoft was overtaken by Google, which now operates the world's most popular operating system (Android, not Windows), browser (Chrome, not Internet Explorer), and online services (Gmail, not Hotmail or Windows Live). Yet what will Google's Metaverse role be? The company's mission is to "organize the world's information and make it universally accessible and useful," but it can access little of the information that exists in virtual worlds, let alone use it. And it has no virtual worlds, virtual world platforms, virtual world engines, or any similar services of its own. Notably, Niantic was originally a subsidiary of Google, but was spun out in 2015. Two years later, Google sold its satellite imaging business to Planet Labs. In 2016, the company began building a cloud game-streaming service, Stadia, which launched at the end of 2019. Earlier that year, Google also announced the Stadia Games and Entertainment division, a "cloud native" content studio. In early 2021, this studio was shut down. In the months that followed, many top Stadia executives, including its general manager, moved to other groups within Google, or exited the company entirely.

We can already see evidence of new disruptors in companies such as Epic Games, Unity, and Roblox Corporation. Though their valuations, revenues, and operational scale are modest compared to GAFAM, they have the player networks, the developer networks, the virtual worlds, and the "virtual plumbing" to be real leaders in the Metaverse. Not only that, but their histories, cultures, and skillsets have refreshingly little in common with the world's current tech titans—even if all of these companies agree that the Metaverse is the future. For much of the past decade and a half, GAFAM has mostly concerned itself with other bets, including streaming TV, social video and live video, cloud-based word processors, and data centers. Nothing is wrong with this focus, but comparatively little attention was paid to video games, least of all to the idea that the best onboard to "the Metaverse" was battle royales, virtual playgrounds for children, or even just game engines. The tech giants' relative disregard of gaming is emblematic of the challenges of preparing for—and predicting—a shift to a new era.

Not long after Mark Zuckerberg acquired Instagram for $1 billion in 2012, the deal was seen as one of the most brilliant acquisitions of the digital era. At the time, the image-sharing service had barely 25 million monthly active users, a dozen employees, and no revenue. A decade later, its estimated value exceeds $500 billion. WhatsApp, which Facebook bought two years later for $20 billion, at which point it had 700 million users, is seen in a similar light. Both are now widely considered not only brilliant acquisitions, but moves that regulators should have blocked on antitrust grounds.

Despite the widespread reverence for Zuckerberg's acquisitions record, neither Facebook nor its competitors acquired Epic, Unity, or Roblox—even though these companies spent most of the last decade valued at the low-single-digit billions—less than a week's profits for most of the GAFAM companies.* Why? The role and potential of each of these companies was simply too uncertain. The video games domain was considered niche at best, fringe at worst. Recall that Neal Stephenson didn't originally envision the category as the only ramp to the Metaverse, either—but by 2011, he was stating that it was and nearly every tech executive in the West had at least heard of, if not played, *Second Life* and *World of Warcraft*.

To Zuckerberg's credit, leaked memos show that in 2015 he pitched his board on acquiring Unity, which had not yet become a unicorn. However, there are no reports of an official bid even though it could've been had cheaply: it wasn't until 2020 that Unity's valuation grew above $10 billion. While Facebook did acquire Oculus VR in 2014, the platform has had fewer lifetime users than Epic, Unity, and Roblox will have in the next 24 hours. This doesn't mean Oculus was a mistake; it may yet be transformative—but Facebook was not limited to a single acquisition (indeed, it has made dozens since). In addition, the ostensible core of Facebook's Metaverse strategy is not Oculus, nor VR

---

* Most of the major Hollywood companies bragged about how they "almost bought Netflix" or "thought about buying Instagram," so it's notable that had any of them bought Epic, Roblox, or Unity, it's likely the acquisition would now be worth more than its parent company.

and AR, but the *Roblox* and *Fortnite*-like *Horizon Worlds* integrated virtual world platform (which is built on Unity). And Roblox has the exact consumers who threaten Facebook's future—not those disengaging with the social network, but those who never even adopted it.

If Facebook is the most aggressive investor in the Metaverse, and Google the most poorly positioned, Amazon sits somewhere in the middle. Amazon Web Services has nearly a third of the cloud infrastructure market and, as discussed throughout this book, the Metaverse will demand unprecedented computing power, data storage, and live services. AWS, in other words, benefits even if other cloud providers take a greater share of future growth. However, Amazon's efforts to build Metaverse-specific content and services have been largely unsuccessful and arguably less of a priority compared to more traditional markets, such as music, podcasting, video, fast fashion, and digital assistants. According to various reports, Amazon has spent hundreds of millions each year on Amazon Game Studios, which focused on Amazon founder Jeff Bezos's goal of making "computationally ridiculous games." However, most of these titles ended up cancelled before release (though not until their development budgets exceeded the lifetime budgets of most hit games). *New World*, released in September 2021, received strong reviews and initial interest (incredibly, it ran out of available AWS servers), but its monthly player count is estimated in the low millions. Another helpful example is *Lost Ark*, which Amazon Game Studios released to acclaim in February 2022. Success is always nice, but *Lost Ark* was not made by AGS, just republished. The title was developed by Smilegate RPG and released in South Korea in 2019, with Amazon striking a deal for English-language territories a year later. More hits are likely to come, but the several billion spent per year on Amazon Music and Amazon Prime Video (and the $8.5 billion acquisition of Hollywood studio MGM) stand in clear contrast. According to some reports, Amazon will spend more on a single season of its *Lord of the Rings* television series than it spends on its gaming studio annually. A similar example comes from Amazon's cloud game-streaming service, Luna, which launched in October 2020, yet found even less of a market than Google Stadia and included almost no free content for subscribers (which again differs from other Amazon content offerings). Four months after Luna launched, the executive who oversaw the division left to become general manager of the Unity Engine. Amazon's effort to build a competitor to Steam has also been unsuccessful, despite the ongoing strength and success of Twitch, the market leader in live video game broadcasting, and the Prime membership program.

Amazon's most noteworthy gaming initiative started in 2015, when it spent a reported $50 million to $70 million to license the CryEngine, a middling independent game engine owned by CryTek, the publisher behind the game *Far Cry*. Over the next few years, Amazon invested hundreds of millions transforming CryEngine into Lumberyard, a would-be competitor to Unreal and Unity, albeit one optimized for AWS. The engine never found much adoption, with the Linux Foundation taking over development in early 2021, renaming it the "Open 3D Engine," and making it free and open-source. Amazon may have more success in AR or VR hardware, but thus far, almost all of its efforts in and around real-time rendering, game production, and game distribution have disappointed.

As I discussed in the hardware and payments chapters, Apple is also an inevitable beneficiary of the Metaverse. Even if regulators unbundle many of its services, the company's hardware, operating system, and app platform will remain a key gateway to the virtual world, which will send billions in high margin revenue its way, and amplify its influence over technical standards and business models. The company is also better positioned than any other to launch lightweight, high-powered, and easy-to-use AR and VR headsets, as well as other wearables, in part due to their ability to richly integrate with its iPhone. However, Apple is not known to be developing its own IVWP, such as *Roblox*, a category of application that might intermediate the company from many virtual world users and developers. Given that Apple lacks much gaming expertise and is also understood to be a hardware-, not a software- or network-focused, company, building a leading IVWP is unlikely.

The most interesting GAFAM company in the Metaverse era may well be Microsoft, one of the leading case studies for displacement in the mobile era. Since the very first Xbox released in 2001, investors and even company executives mused as to whether its gaming division was essential, or a distraction. Three months after Satya Nadella took over as CEO from Steve Ballmer, company founder and chairman Bill Gates said he would "absolutely" support Nadella if he wanted to spin off Xbox, "But we're going to have an overall gaming strategy, so it's not as obvious as you might think." The first multi-billion-dollar acquisition Nadella made was for *Minecraft*—and in a move that now seems obvious but was unconventional at the time, opted against making the title exclusive to its Xbox and Windows platforms (or even better on them). Furthermore, the title's playerbase has grown more than 500% since acquisition, from 25 million monthly users to 150 million, making it the second most popular real-time rendered 3D virtual world globally.

As we know, gaming experiences now sit at the forefront of the industry—including at Microsoft. Recall that *Microsoft Flight Simulator* is a marvel of both technology and collaboration. Though Xbox Game Studios developed and published the title, it was built in partnership with Bing Maps and leveraged data from OpenStreetMaps, a collaborative and free-to-use online geographic, with Azure's artificial intelligence bringing this data together into 3D visualizations, powering real-time weather, and supporting cloud data streaming. The Xbox division also has its own hardware suite, the most popular cloud game-streaming service in the world, a fleet of first-party game studios, and a handful of proprietary engines. Although HoloLens is run by the Azure AI division, its adjacency to gaming is obvious. In January 2022, Microsoft agreed to buy Activision Blizzard, the largest independent game publisher outside of China, for $75 billion (the largest acquisition in GAFAM history). In announcing the deal, Microsoft said that "[Activision Blizzard] will accelerate the growth in Microsoft's gaming business across mobile, PC, console and cloud and will provide building blocks for the metaverse."[1]

In many ways, Nadella's approach to *Minecraft* embodied his overall transformation of Microsoft. No longer would the company's products be designed for (or even optimized to work with) its own operating systems, hardware, technology stack, or services. Instead, it would be platform agnostic, supporting as many platforms as possible. This is how Microsoft was able to grow despite losing its hegemony over computing operating systems—the digital world grew more than Microsoft's share contracted. The same philosophy positions the company well for the Metaverse.

Sony, which was founded in 1946, is another intriguing conglomerate. By revenue, Sony Interactive Entertainment (SIE) is the largest gaming company in the world, with this business spanning proprietary hardware and games, as well as third-party publishing and distribution. SIE also operates the world's second-largest paid gaming network (PlayStation Network), the third-largest cloud game-streaming subscription service (PSNow), and several high-fidelity game engines. The company's portfolio of original games, such as *The Last of Us*, *God of War*, and *Horizon Zero Dawn*, are considered among the most vivid and creative in industry history. The PlayStation is also the top-selling console of the fifth, sixth, eighth, and ninth console generations, and will launch its PS VR2 platform in 2022. Sony Pictures, meanwhile, is the largest movie studio by revenue, as well as the largest independent TV/film studio overall. Sony's semiconductor division is also the world leader in image sensors, with nearly 50% market share (Apple is a top customer), while its Imageworks division is a top visual effects and computer animation studio. Sony's Hawk-Eye is a computer vision system used by numerous professional sports leagues globally to aid officiating through 3D simulations and playblack (the football club Manchester City is also deploying the technology to create a live digital twin of its stadium, players, and fans during a match). Sony Music is the second-largest music label by revenue (Travis Scott is a Sony Music artist), while Crunchyroll and Funimation provide Sony with the world's largest anime streaming service. It is impossible to review Sony's assets and creativity capabilities and see anything other than enormous potential as the Metaverse emerges. However, many challenges remain.

Sony's games are almost always PlayStation-only, and SIE has had limited success producing hit mobile, cross-platform, or multiplayer games. Though strong in gaming hardware and content, Sony is typically viewed as a laggard in online services, and has no leadership in compute and networking infrastructure, or virtual production. And despite Japan's strength in semiconductors, the country has not produced any major contenders in this area—meaning Sony's shift to the Metaverse will likely require the use of GAFAM services and products.*

In 2020, Sony released *Dreams*, a powerful IVWP that the company seeded with many professionally produced games, but failed to attract many users or developers. Many critics argued that *Dreams* was always doomed and reflects Sony's inexperience with UGC platforms. Unlike most IVWPs, *Dreams* was not free-to-play, but cost $40. In addition, the title did not offer developers any cut of revenues, and was limited to PlayStation consoles, whereas competing IVWPs were playable on billions of devices worldwide.[†]

Compared to GAFAM, Sony reaches a fraction of users, employs few engineers, and its annual R&D budget is outspent in months or even weeks. For decades the company has been a case study for missed opportunities. Though Sony was the global market leader in portable music devices through the Walkman, and owned the second-largest music label, it was Apple that revolutionized digital music. Despite the company's strength in consumer electronics, smartphones, and gaming, it was also squeezed out of the mobile phone business, and altogether missed the connected TV device category. While Sony was the only Hollywood giant without a legacy TV business to protect,

---

* In May 2019, Sony announced a "strategic partnership" with Microsoft to use its Azure data centers for cloud gaming, among other content streaming services. In February 2020, the head of Xbox said, "When you talk about Nintendo and Sony, we have a ton of respect for them, but we see Amazon and Google as the main competitors going forward. . . . That's not to disrespect Nintendo and Sony, but the traditional gaming companies are somewhat out of position. I guess they could try to re-create Azure, but we've invested tens of billions of dollars in cloud over the years." (Seth Schiesel," Why Big Tech Is Betting Big on Gaming in 2020," *Protocol*, February 5, 2020, https://www.protocol.com/tech-gaming-amazon-facebook-microsoft.)

† Limiting *Dreams* to PlayStation devices is partly why the title was so technically powerful, as mobile devices are obviously less capable computing devices. But by originally architecting the IVWP for its own high-end device, Sony has also made it more difficult to ever expand the title to other platforms.

and launched its streaming service Crackle the same year Netflix pivoted from DVDs, it failed to capitalize on the opportunity. To lead in the Metaverse, Sony will need not just considerable innovation, but unprecedented cross-division collaboration—the sort that challenges even the most integrated of companies. And at the same time, the company will need to move outside of its own tightly integrated ecosystems, such as PlayStation, and connect into third party platforms too.

Then there's Nvidia, a company built over 30 years specifically for the era of graphics-based computing. Alongside major processor and chip companies such as Intel and AMD, Nvidia will benefit from any incremental demand for compute. The high-end GPUs and CPUs inside our devices, as well as the data centers of Amazon, Google, and Microsoft, typically come from these providers. Nvidia, though, aspires for far more. For example, the company's GeForce Now cloud game-streaming service is the second most popular in the world, several times the size of Sony, orders of magnitude larger than Amazon's Luna or Google's Stadia, and half that of market leader Microsoft. Its Omniverse platform, meanwhile, is pioneering 3D standards, facilitating the interoperation of disparate engines, objects, and simulations, and may yet become a sort of *Roblox* for "digital twins" and the real world. We may never wear Nvidia-branded headsets nor play Nvidia-published games, but at least in 2022, it looks likely that we live in a Metaverse powered in large part by Nvidia.

The danger in assessing the preparedness of today's leaders for tomorrow's future is that they always look prepared. And that's because they are—they have cash, technology, users, engineers, patents, relationships, and more. Yet we know that some of these companies will falter, often because of these many advantages (some of which will turn out to be encumbrances). In time, it will become clear that many of the leaders in the Metaverse weren't even mentioned in this book—perhaps because they were too small to be of note, or unknown to its author. Some hadn't even been created let alone thought up. An entire generation of *Roblox*-natives is only now on the cusp on adulthood, and it's likely they, not Silicon Valley, will create the first great game that has thousands (or tens of thousands) of concurrent users, or blockchain-based IVWP. Whether motivated by Web3 principles, emboldened by the trillion-dollar opportunity the Metaverse provides, or simply unable to sell to GAFAM due to regulatory scrutiny, these founders will ultimately displace at least one member of the GAFAM five.

## Why Trust Matters More Than Ever

Regardless of which companies come to dominate, the most likely outcome is indeed that a handful of vertically and horizontally integrated platforms collect a significant share of total time, content, data, and revenues in the Metaverse. This doesn't mean a majority of any of these resources—recall that GAFAM represents less than 10% of total digital revenues in 2021—but enough to collectively shape the economy of the Metaverse and the behaviors of its users, as well as the economy of the real world and its citizens.

All business, and especially software-based business, benefits from feedback loops—more data leads to better recommendations, more users means stickier users and more advertisers, greater revenues enable more licensing spend, larger investment budgets attract more talent. This general point doesn't change in a blockchain future for the same reason audiences still converged on a handful of websites and portals, such as Yahoo or AOL in the 1990s, even though millions of other sites were available. Habits are themselves sticky, which is part of the reason even blockchain dapps are valued in the billions by venture capitalists—even though their authority over their users or their data is marginal compared to the "Web 2.0" era.

To many, however, the real war for the Metaverse is not between major corporations, or between these companies and the start-ups that hope to displace them. Instead, the war is between "centralization" and "decentralization." Of course, this frame is imperfect because neither side can "win." What matters is where the Metaverse falls between the two poles, why, and how its position shifts over time. When Apple launched its closed mobile ecosystem in 2007, it was betting against conventional wisdom. The success of this bet has doubtlessly led to a larger and more mature digital, and especially mobile, economy, while also creating the most valuable and profitable company and product in history. But 15 years later, with Apple's share of US personal computers now up from less than 2% to more than two-thirds (with its share of software sales sitting closer to three-fourths), Apple's dominance now holds back the entire industry by depriving developers and consumers with much of a choice. While testifying as part of Epic Games' lawsuit against the company, Apple CEO Tim Cook told the judge that even allowing developers to have an in-app link that would send them to alternative payment solutions would mean "essentially [giving] up the total return on our IP."[2] No next-generation internet should be so constrained by such policies. And yet, *Roblox*, the most popular "proto-Metaverse" thus far, thrives for many of the same reasons that Apple's iOS did: tight control over as much of its experience as possible, including forced bundling of content, distribution, payments, account systems, virtual goods, and more.

With this in mind, we should acknowledge that the growth of the Metaverse benefits from both decentralization *and* centralization— just like the real world. And again, just like the real world, the middle ground isn't a fixed point, nor even a knowable one, let alone one that's agreed upon. But there are some obvious policy approaches that follow if most companies, developers, and users accept the basic point that it cannot be one or the other.

For example, Epic Games' Unreal license to developers is written in a way that gives licensees indefinite rights to a specific Unreal Engine build. Epic can still change its license for subsequent builds and updates, such as 4.13 and especially 5.0 or 6.0—and giving away such a right would be financially impractical and probably harmful to developers as a result. But the result of this policy is that developers need not worry that by choosing to use Unreal, they're forever reliant upon the whims, desires, and leadership of Epic (after all, there's no rent control board in the Metaverse, nor appeals court). And as Unreal's license allows developers nearly free rein on customizations and third-party integrations, developers can choose not to use future updates and instead build their own in lieu of whatever Epic adds in 4.13, 4.14, 5.0, and beyond.

In 2021, Epic made another important modification to its Unreal license: it gave up the right to terminate that license, even in the instance where a developer had failed to make an outstanding payment or violated the agreement outright. Instead, Epic would need to take their customer to court in order to mandate payment or win an injunction that would allow them to suspend support. This made it harder, slower, and costlier for Epic to enforce its rules, but the policy is designed to build trust with developers, and Epic hopes it will be good business overall. Imagine if your landlord could lock you out of your apartment at any point by arguing that you violated your rent agreement, or you missed a payment by a day—or even 60 days. This would not only be bad for your psychological health, but it would also discourage renting and, well, living in the city in the first place. In the Metaverse, tenants can be locked out, or permanently banned without much cause, and their possessions permanently revoked. The tech libertarian answer is decentralization, likely through blockchain. Another, not mutually exclusive, answer is to extend the legal systems of the "real world" to reflect the materiality of the immaterial. Tim Sweeney argues that no one benefits from "powerful companies [having] the ability to act as judge, jury, and executioner," able to stop a business from "building products," "distributing their product," or servicing "customer relationships."

My great hope for the Metaverse is that it will produce a "race to trust." To attract developers, the major platforms are investing billions to make it easier, cheaper, and faster to build better and more profitable virtual goods, spaces, and worlds. But they're also showing a renewed interest in proving—through policy—that they deserve to be a partner, not just a publisher or platform. This has always been a good business strategy, but the enormity of the investment required to build the Metaverse, and the trust it requires from developers, has placed this strategy front and center.

In April 2021, Microsoft announced that games sold on its PC Windows Store would pay only a 12% fee, rather than the customary 30% (which remained in place on Xbox), and that Xbox users could play free-to-play games without needing to subscribe to the console's Xbox Live service. Two months later, this policy was revised so that non-gaming apps could use their own billing solution, rather than Microsoft's, and therefore pay only the 2%–3% charged by an underlying payment rail, such as those of Visa or PayPal. By September, Xbox announced that its Edge browser had been updated to "modern web standards," enabling users to play cloud game-streaming services owned by Xbox's competitors, such as Google's Stadia and Nvidia's GeForce Now, from the device and without using Microsoft's store or live services.

Microsoft's most significant policy change occurred in February 2022, when the company announced a new, fourteen-point policy platform for its Windows operating system, and the "next-generation marketplaces [the company] builds for games." This included a commitment to support third party payment solutions and app stores (and not disadvantage developers who choose to use them), the right for users to set these alternatives as default options, and the right for developers to directly communicate with the end-user (even if the point of that communication is to tell the user they can get better pricing or service by cutting out Microsoft's store or services suite). Crucially, Microsoft stated that not all of these principles would "apply immediately and wholesale to the current Xbox console store," as Xbox hardware was designed to sell at a loss and generate a cumulative profit through software sold by Microsoft's proprietary store. However, Microsoft said "we recognize that we will need to adapt our business model even for the store on the Xbox console. . . . We're committed to closing the gap on the remaining principles over time."[3]

When he was unveiling Facebook's Metaverse strategy in October 2021, Mark Zuckerberg was clear about the need to "maximize the economy of the Metaverse" and support developers. To this end, Zuckerberg made a series of policy commitments which, at least based on the approaches taken by other software platforms today, benefit developers by marginalizing the power and profits of Facebook's VR (and also forthcoming) AR devices. For example, Zuckerberg said that while Facebook's devices would continue to be sold at or below cost, (similar to consoles but unlike smartphones), the company would allow users to download apps directly from the developer or even through competing app stores. He also announced that Oculus devices would no longer require a Facebook account (which had become a new policy in August 2020), and would continue to use WebXR, an open-source API collection for browser-based AR and VR apps, and OpenXR, an open-source API collection for installed AR and VR apps, rather than produce (let alone require) their own proprietary API suite. Recall from Chapter 10 that almost all other computing platforms either block rich browser-based rendering, and/or require the use of a proprietary API collections.

In the weeks that followed, Facebook also began to enable several APIs and integrations with competing platforms that had once been supported, but been closed for several years. One of the most noteworthy examples involved the ability to post an Instagram link to Twitter whereby the relevant Instagram photo would display inside a tweet. Instagram offered this API not long after it launched in 2010, but removed it only eight months after the company was acquired by Facebook in 2012.

It's easy to be cynical about the maneuvers of Microsoft, Facebook, and other "Web 2.0" giants. In May 2020, Microsoft's president Brad Smith said that the company had been "on the wrong side of history" when it came to open-source software, then in February 2022, he publicly endorsed a bill passed by the US Senate which would require Apple and Google to open up their mobile operating systems to third-party app stores and payment services (he said the "important" legislation "would promote competition, and ensure fairness and innovation").[4]

Had the company thrived in mobile, as Apple and Google did, rather than been displaced by those companies, or had Xbox ranked first among consoles, not last, Microsoft may not have changed its view. If Facebook had its own operating system, rather than been stymied by its lack thereof, would it be so relaxed about sideloading? If it weren't so late to building a popular gaming platform, would Facebook have really wanted to rely on OpenXR and WebXR? These points are fair, but they also ignore the many genuine (if undesirable) lessons learned by platform makers and developers over the past decades. And these two groups aren't the only constituencies who are smarter today than in Y2K.

As the "trustless" and "permissionless" nature of blockchain programming suggests, much of the Web3 movement stems from a dissatisfaction with the last 20 years of digital apps, platforms, and ecosystems. Yes, we received many great services for free during "Web 2.0," such as Google Maps and Instagram, and many careers and businesses have been built on top of and through these services. Still, many believe the exchange was not a fair one. In return for "free service," users provided these services with "free data" that have been used to build companies worth hundreds of billions or even trillions of dollars. Worse still, these companies effectively own data in perpetuity, which in turn makes it difficult for the user who generated the data to use it elsewhere. Amazon's recommendations, for example, are so powerful because they're based on years of prior searches and purchases—but as a result, even with equivalent inventory, lower prices, and similar technology, Walmart (or other "upstarts") will always have a harder path toward making an Amazon customer happy. Many people argue that Amazon should therefore have to provide users with the right to export their history and take it to competing sites. Instagram users can technically export all their photos into a downloadable zip file, then upload them to a competing service, but it's not an easy process, and there's no way to carry over each photo's likes and comments. Overall, many people have also come to believe that companies built "off of their data" have dramatically worsened the real world, adversely affecting the psychological and emotional lives of those who use their services. A good portion of the reaction to Zuckerberg's announcement of the name change to Meta consisted of derision. Why should a company like Facebook have even more reach into our lives? Hasn't big tech already created too much of the dystopias described by Gibson, Stephenson, and Cline?

It should then come as no surprise that the terms "Web3" and "Metaverse" have been conflated. If one disagrees with the philosophy and arc of Web 2.0, then it's terrifying to think of the power bestowed upon the tech behemoths when they operate a parallel plane of existence—when the "atoms" of the virtual universe are written, executed, and transmitted by for-profit corporation. Envisioning the Metaverse as dystopic solely because the term and many of its inspirations come from dystopic science fiction is misguided, but there's a reason those who control these fictional universes (the Matrix, the Metaverse, the Oasis) tend to use it for ill: their power is absolute, and absolute power corrupts. Recall Sweeney's warning: "If one central company gains control of the [Metaverse], they will become more powerful than any government and be a god on Earth."

All of this leads to one of the most important aspects of any serious discussion of the Metaverse: how it will affect the world around us and the policies we'll need to shape its impact.

# Chapter 15

# METAVERSAL EXISTENCE

THE DIGITAL ERA HAS IMPROVED MANY ASPECTS of our lives. There has never been greater access to information, nor a time when so much of the information available to us was free. Many marginalized groups and individuals now have large and unstoppable digital megaphones in their hands. Those who are physically far apart can feel closer to each other. Art has never been so easy to find, nor so many artists paid for their work.

Yet decades after the Internet Protocol Suite was established, we as a society still contend with numerous challenges in our online lives: misinformation, manipulation, and radicalization; harassment and abuse; limited data rights; poor data security; the arguably constraining and inflaming role of algorithms and personalization; general unhappiness as a result of online engagement; immense platform power amid toothless regulation; among many others. These problems have mostly grown with time.

Though they are delivered, facilitated, or exacerbated by technology, the challenges we face in the mobile era are human and societal problems at their core. As more people, time, and spending go online, more of our problems go online, too. Facebook has tens of thousands of content moderators; if hiring more moderators would solve for harassment, misinformation, and other ills on the platform, no one would be more motivated to do so than Mark Zuckerberg. And yet the tech world, including hundreds of millions if not billions of everyday users—think of all the individual creators in *Roblox*, for example—are pressing on to the "next internet."

The very idea of the Metaverse means that more of our lives, labor, leisure, time, spending, wealth, happiness, and relationships will go online. Actually, they will *exist* online, rather than just be put online like a Facebook post or Instagram upload, or aided by digital devices and software, as a Google Search or iMessage might. Many of the benefits of the internet will grow as a result, but this fact will also exacerbate our great and unsolved socio-technological challenges. These will also permutate, making it difficult to simply reapply the lessons learned from the past 15 years of the social and mobile internet.

In the mid-2010s, the militant Sunni group the Islamic State, commonly referred to as ISIS, used social media to radicalize foreign nationals who would then visit Syria for training. This led to many "red flags" for those with travel records that included time in Syria, among other Middle Eastern nations, as various countries grappled with the threat of their citizens becoming combatants. Rich real-time rendered 3D virtual worlds will assuredly make radicalization easierand offer better training to those who never leave their native country (and for some of the same reasons that remote education will improve). At the same time, the Metaverse may make learning about and tracking people through their digital activity even easier, with perhaps many more peo ple ending up on government lists or under government surveillance.

Misinformation and election tampering will likely increase, making our current-day complications of out-of-context sound bites, trolling tweets, and faulty scientific claims feel quaint. Decentralization, often seen as the solution to many of the problems created by the tech giants, will also make moderation more difficult, malcontents harder to stop, and illicit fundraising far less difficult. Even when limited primarily to text, photos, and videos, harassment has been a seemingly unstoppable blight in the digital world—one that has already ruined many lives and harmed many more. There are several hypothesized strategies to minimize "Metaverse abuse." For example, users may need to give other users explicit levels of permission to interact in given spaces (e.g., for motion capture, the ability to interact via haptics, etc.), and platforms will also automatically block certain capabilities ("no-touch zones"). However, novel forms of harassment will doubtlessly emerge. We are right to be terrified by what "revenge porn" might look like in the Metaverse, powered by high-fidelity avatars, deepfakes, synthetic voice construction, motion capture, and other emergent virtual and physical technologies.

The question of data rights and usage is more abstract, but just as fraught. There is not only the issue of private corporations and governments accessing personal data but also more fundamental issues, such as whether users understand what they're sharing. Are they valuing it appropriately? What obligations does a platform have to give data back to that user? Should a free service have to offer users the option to "buy out" data collection, and if so, how would this be valued? We do not have perfect answers to these questions right now, nor ways to find them. But the Metaverse will mean placing more data and more important information online. It will also mean sharing this data with countless third parties, while also enabling these parties to modify the data. How is this new process managed securely? Who manages it? What is the recourse for mistakes, failures, losses, and breaches? For that matter, who should own virtual data? Should a business that spends millions developing inside *Roblox* have a right to what they built? A right to take it elsewhere? Does a user who bought land or goods inside of *Roblox* hold that right? Should they?

The Metaverse will further redefine the nature of work and labor markets. Right now, the majority of offshored jobs are menial and audio-only, such as technical support and bill collection. The gig economy, meanwhile, often takes place in person, but is not altogether dissimilar: ridesharing, housecleaning, dog walking. This will change as virtual worlds, volumetric displays, live-motion capture, and haptic sensors improve. A blackjack dealer need not live anywhere near Las Vegas, or even in the United States, to work at a casino's virtual twin. The world's best tutors (and sex workers) will program and then participate in hourly experiences. A retail store employee might "call in" from thousands of miles away—and be better off for it. Rather than wander the store waiting for a customer, they'll come when a customer consultation is needed and, through tracking and projection cameras, they will be able to counsel on where, say, alternative sizes or tailoring might help.

But what does the Metaverse mean for hiring rights and minimum-wage laws? Can a Mirror instructor live in Lima? Can a blackjack dealer be in Bangalore? And if they can, how does that affect the supply of in-person labor (and the prices paid for in-person labor)? These are not altogether new questions, but they will become more significant if the Metaverse becomes a multi-trillion-dollar part of the world economy (or, as Jensen Huang expects, more than half of it). Among the darkest visions of the future is one in which the Metaverse is a virtual playground where the impossible is possible, but it is powered by toiling "third-world" laborers for the sake of "first-world" joys.

There is also the question of identity in the virtual world. While modern society grapples with questions of cultural appropriation and the ethics of clothing and hairstyles, we're confronting the tension between using avatars to reveal a different, and potentially truer, version of ourselves, and the need to reproduce it faithfully. It is acceptable for a white man's avatar to be that of an Aboriginal woman? Does the realism of the avatar matter in answering that question? Or for that matter, whether it's made of (virtual) organic material or metal?

Questions of identity online have been raised recently around the Cryptopunks NFT collection, for example. Recall that there are 10,000 of these algorithmically generated, 24 × 24–pixel, 2D avatar "cryptopunks," all of which are minted to the Ethereum blockchain and are typically used as profile pictures on various social networks. On any given day, it's likely that the cheapest Cryptopunks listed for sale are those with dark pigmentations. Some believe this price dynamic is an obvious manifestation of racism. Others argue that it reflects the belief that it isn't appropriate for white members of the cryptocurrency community to use these Cryptopunks. Those who hold this view also assert that it's not even appropriate for white people to own them. If so, the price discount reflects the fact that the number of white Cryptopunks are disproportionately low compared to the composition of the United States, where most Cryptopunks are bought and sold, and the Cryptocommunity overall. Thus it's not that the prices for the "non-white" Cryptopunks are low, but that "white" Cryptopunks are too scarce. One position is that perhaps the "discount" on the former is positive—it makes these would-be avatars and supposed membership cards more affordable to those who have less wealth in general.

Other concerns include the "digital divide" and "virtual isolation," though these appear easier to address. A decade ago, some worried that the adoption of superpowered mobile devices—most of which cost hundreds of dollars more than a "dumbphone"—would exacerbate inequality. The most frequently used example was that of iPads in education. What would happen if some students couldn't afford the device and had to rely on "analog," out-of-date, and unpersonalized textbooks, while their wealthy peers (whether they sat beside them or in exclusive private schools miles away) took advantage of digital and dynamically updated textbooks? Such worries have been assuaged by the rapidly declining cost of these devices, as well as their ever-expanding utility. In 2022, a new iPad can be purchased for less than $250—making it cheaper than most PCs, even though it's considerably more capable. The most expensive iPhone costs three times as much as the 2007 original, but the most affordable iPhone sold by Apple is 20% cheaper (40% cheaper after adjusting for inflation) and offers more than a hundred times the computing power. And none of these devices need to be bought for the classroom; most students already own one. This is the arc of most consumer electronics: they begin as a toy for the wealthy, but early sales enable more investment, which leads to cost improvements, which drive greater sales, which facilitate greater production efficiency, leading to lower prices, and so on. VR and AR headsets will be no different.

It is natural to worry about a future in which no one goes outside and spends their existence strapped to a VR headset. Yet such fears tend to lack context. In the United States, for example, nearly 300 million people watch an average of five and a half hours of video per day (or 1.5 billion hours in total). We also tend to watch video alone, on the couch or in bed, and none of it is social. As those in Hollywood often boast, this content is passively consumed (in industry jargon, it is "lean-back entertainment"). Shifting any of this time to social, interactive, and more engaged entertainment is likely a positive outcome, not a negative one, even if we're all still indoors. This is particularly true for the elderly. The average senior in the United States spends seven and a half hours per day watching TV. Few among us dreamed of retirement and long life in order to spend half of each of our remaining days watching TV. The Metaverse may offer no substitute for actually sailing in the Caribbean, but manning a virtual sailboat alongside old friends is likely to come pretty close and offer all sorts of digital-only perks—and beat watching midday Fox News or MSNBC.

## Governing the Metaverse

For the same reasons the Metaverse is so disruptive—it's unpredictable, recursive, and still vague—it is impossible to know what problems will emerge, how best to solve those which already exist, and how best to steer it. But as voters, users, developers, and consumers, we have agency. Not just over our virtual avatars as they navigate virtual space, but over the broader issues surrounding who builds the Metaverse, how, and upon which philosophies.

As a Canadian, I probably believe in a larger governmental role in the Metaverse than many others—even though I've spent a good portion of my life thinking and writing and talking about what some consider a free-market capitalist's dream. What is clear, however, is that one of the larger challenges facing the Metaverse is that it lacks governing bodies beyond virtual world platform operators and service providers. By now you should be convinced that these groups aren't enough to create a healthy Metaverse.

Recall the importance of the Internet Engineering Task Force. This body was originally established by the US federal government to steer voluntary internet standards, especially of TCP/IP. Without the

IETF, and other nonprofit bodies, some of which were created by the Department of Defense, we would not have the internet as we know it. Instead, it would likely be a smaller, more controlled, and less vibrant internet—or perhaps one of several different "nets."

The IETF is largely unknown to younger generations, even though its work continues to this day. But the organization's mostly behind-the-scenes contributions are one of the reasons why many believe Western nations are incapable of effective tech regulation or oversight. I'm not referring to antitrust, although that is a pressing issue. Rather, I mean the idea of a role for government in the development of technology. In truth, the apparent divide between government and technology is a relatively recent problem. Throughout the 20th century, governments proved more than capable of steering new technologies, from telecommunications to railroads, oil, and financial services—and, obviously, the internet. It's only in the last 15 years or so that they have fallen short. The Metaverse presents an opportunity not just for users, developers, and platforms, but for new rules, standards, and governing bodies, as well as new expectations for those governing bodies.

What should these policies look like? Let me start with a transparent confession. As these questions encompass ethics, human rights, and annals of case law, I am deliberately cautious and modest. There are clear social justice issues that go beyond many of those detailed throughout this book, such as the devices used to access the Metaverse (and their cost), the quality of the experience these devices provide, and the platform fees collected. I am aware of these, and aware of others' authority to speak to them with greater clarity. Rather, I will provide a framework that reflects my own areas of expertise and elaborates on issues raised in the book's previous chapters.

In 2022, many governments, including those of the United States, the European Union, South Korea, Japan, and India, are focused on whether Apple and Google should have unilateral control over in-app billing policies and the right to block competing payment services or disintermediate other payment rails (for example, ACH and wire). Dismantling the hegemony of Apple and Google would be a good start and quickly increase developer margins and/or reduce consumer prices, enable new businesses and business models to thrive, and eliminate the inconsistent commissions which encourage developers to focus on physical goods or advertising rather than virtual experiences and consumer spending. But, as we've seen, payments are just one of many levers that platforms use to assert control over developers, users, and potential competitors. Apple's and Google's goal is to maximize their respective shares of online revenues. Accordingly, regulators should force platforms to unbundle identity, software distribution, APIs, and entitlements from their hardware and operating systems. For the Metaverse and the digital economy to thrive, users must be able to "own" their online identity and the software they purchase. Users must also be able to choose how they install and pay for this software, while developers need to be free to decide how their software is distributed on a given platform. Ultimately, these two groups should be able to determine which standards and emerging technologies are best, irrespective of the preferences of the company whose operating system runs the resulting code. Unbundling would force OS-centric companies to compete more clearly on the merits of their individual offerings.

We also need greater protections for the developers who build on independent game engines, integrated virtual worlds, and app stores. Sweeney's approach to Unreal's license to developers is the right one—handing control over the termination of that license to court processes, rather than internal corporate ones. However, for-profit corporations should not be the only groups who decide where their de facto laws end and legislative/judicial processes begin. We cannot count on their altruism, even if, as in the case of Epic, that "altruism" is linked to better business practices. Critically, unless new laws are written specifically for virtual assets, virtual tenancy, and virtual communities, it's likely that those designed for the era of physical goods, physical malls, and physical infrastructure will end up misapplied and exploited. If the economy of the Metaverse will one day rival that of the physical world, then governments need to take the jobs, business transactions, and consumer rights inside of it just as seriously.

A good place to start would be enacting policies regarding how and to what extent IVWPs should be required to support developers who want to export the environments, assets, and experiences they've created. This is a relatively new problem for regulators. On the current internet, nearly every online "unit of content," from a photo to text, audio file, or video, can be transferred between social platforms, databases, cloud providers, content management systems, web domains, hosting companies, and more. Code is mostly transferrable, too. Despite this, it's obvious that content-focused online platforms aren't struggling to build a multi-billion-dollar (or trillion-dollar) business. These companies don't need to "own" a user's content in order to produce a flywheel based around its consumption. YouTube is the perfect example. It's easy for a YouTuber to decamp for another online video service—and take their entire library with them—but they stay because YouTube offers content creators greater reach and, typically, higher incomes.

It is the very fact that a YouTuber can so easily leave for Instagram, Facebook, Twitch, or Amazon has led many other platforms to try to poach YouTube's content creators. This, in turn, pushes YouTube to innovate, work harder to satisfy its content creators, and be a more responsible platform overall. Similarly, the fact that a Snapchat creator can just as easily publish their content to all social services, from Instagram to TikTok, YouTube, and Facebook, means that they can expand their audience without multiplying their production budgets. If a platform, like YouTube, wants a given creator to be exclusive, the platform has to pay for this exclusivity, rather than rely on the fact that it's too difficult and costly for a creator to operate on multiple platforms. There is a reason why every social network has shifted over time to original programming, revenue guarantees, and creator funds.

Unfortunately, the dynamics that apply to "2D" content networks don't easily carry over to IVWP. Most of the content made on YouTube or Snapchat isn't produced using those platforms' tools. Instead, it's produced with independent applications, such as Apple's Camera app, or Adobe's Photoshop and Premiere Pro. Even when content is made on a social platform, such as a Snapchat Story, which uses Snap's filters, the content is typically easy to export (and to use again on

Instagram) because it is just a photo. Conversely, the content made for an IVWP is mostly made in that IVWP. It cannot be easily exported, or repurposed—and there are no available "hacks" similar to using an iPhone's "screenshot" function to grab a Snapchat Story. As such, content made on *Roblox* is essentially *Roblox*-only. And unlike a YouTube video or Snapchat Story, *Roblox* content is not ephemeral (like a live stream), nor is it ever intended to be catalogued (as is the case with a YouTuber's vlogs). Instead, it is intended to be continuously updated.

The consequences of these differences are profound. If a developer wants to operate across multiple IVWPs, they must rebuild nearly every part of their experiences—an investment that produces no value to users and wastes time and money. In many cases, a developer won't even bother, thereby limiting their reach and concentrating their reliance on a single platform. The more a developer invests in a given IVWP, the harder it becomes for them to ever leave—not only will they need to reacquire their customers, they'll have to rebuild from scratch. Thus, developers will be less likely to support new IVWPs that might offer superior functionality, economics, or growth potential—and existing IVWPs will face less pressure to improve. Over time, dominant IVWPs might even "rent-seek." Over the past decade, most of the major platforms have been criticized for such behaviors. For example, many brands argue that changes made to Facebook's Newsfeed algorithm effectively forced them to buy ads in order to reach the very Facebook users who had voluntarily "liked" their Facebook pages. In 2020, Apple revised its App Store policy such that, with a few exceptions, any iOS app that used third-party identity systems (for instance, log-in using your Facebook or Gmail account) would also need to support the Apple account system.

Some IVWPs do support selective exports. *Roblox* enables users to take models produced in *Roblox* and bring them into Blender using the OBJ file format. But as we've seen throughout this book, taking data out of a system doesn't mean it will then be usable data. Even if it is usable, the process to make it so isn't necessarily easy (just try downloading your Facebook data and importing it to Snapchat) and it is up to the discretion of the platform (recall Instagram shutting down the API used to share posts on Twitter). In this sense, governments have both an obligation to regulate as well as an opportunity to shape the standards of the Metaverse. By setting the export conventions, file types, and data structures for IVWPs, regulators would also be informing the import conventions, file types, and data structures of any platform that wants to access this data. Ultimately, we should want it to be as easy as possible to take a virtual immersive educational environment or AR playground from one platform to another—as easy as it is to move a blog or a newsletter. Granted, this goal isn't fully attainable—3D worlds and logic aren't as simple as HTML or spreadsheets. But it should be our target and matters far more than the establishment of standardized charging ports.

It may seem unfair that the companies that helped build the mobile era (such as Apple and Android), as well as those helping to found the era of the Metaverse (namely but not exclusively *Roblox* and *Minecraft*), should be forced to relinquish control over their ecosystems and let competitors profit from their success. After all, it is the rich integration between these platforms' many services and technologies that made them so successful. Such regulations, however, would be best thought of as a reflection of and response to this success—and of what's needed to maintain a market that is collectively prosperous and able to produce new leaders. When Apple revised its cloud gaming policies in September 2020, *The Verge* wrote that "Arguing over whether Apple's guidelines did or didn't include a thing is kind of pointless, though, because Apple has ultimate authority. The company can interpret the guidelines however it chooses, enforce them when it wants, and change them at will."[1] This is not a reliable foundation for the digital economy, let alone the Metaverse.

Beyond regulating the major platforms, we can identify other obvious laws and policy changes that will help produce a healthy Metaverse. Smart contracts and DAOs should be legally recognized. Even if these conventions, and blockchains overall, do not endure, legal status will inspire more entrepreneurship, protect many from exploitation, and lead to wider usage and participation. Economies flourish when this occurs. Another clear opportunity is the expansion of so-called KYC (Know Your Customer) regulations for cryptocurrency investments, wallets, content, and transactions. These regulations would require platforms such as OpenSea, Dapper Labs, and other major blockchain-based games to validate the identity and legal status of customers, while also providing requisite filings to governments, tax bodies, and securities agencies. The nature of blockchains is such that KYC requirements cannot reach everything "crypto"—not unlike the fact that neither the IRS nor police can monitor all cash transactions. But if nearly all mainstream services, marketplaces, and contract platforms mandate this information, then most transactions will occur under such requirements and those which don't will be discounted due to the perceived risk of a scam (just as most would rather use eBay and buy from verified sellers than purchase through an unbranded marketplace and from an anonymous account).

One final proposal is that government should take a far more serious approach to data collection, usage, rights, and penalties. The amount of information that Metaverse-focused platforms will actively and passively generate, collect, and process will be extraordinary. The data will span the dimensions of your bedroom, the detail of your retinas, the facial expressions of your newborn, your job performance and compensation, where you've been, for how long, and probably why. Nearly everything you say and do will be captured by one camera or microphone or another, then sometimes placed in a virtual twin owned by a private company that shares it with many more. Today, what's permissible is often up to the developer or the operating system that runs the developer's application—and only lightly understood by the user. Regulators would do well to lead with, and then occasionally expand, what's allowed, rather than merely respond to unforeseen consequences. Including under "what's allowed" should be the user's right to request the deletion of data, or to download it and easily upload it elsewhere. This is yet another area where governments can, and should, dictate the standards of the Metaverse.

Equally important is how corporations demonstrate their ability to secure privileged information, and how they are punished when they fail to do so. The US Federal Reserve routinely "stress tests" banks to ensure they can withstand economic shocks, market crashes, and mass withdrawals, while also holding executives individually liable for corporate negligence or financial misstatements. Primitive versions of such oversight mechanisms exist today for user data, but they're mostly informal inquiries, rather than formalized processes—and big tech is unlikely to volunteer for audits. Fines for data breaches and losses are particularly toothless. In 2017, American consumer credit reporting agency Equifax revealed that foreign hackers had been illegally accessing its systems for more than four months, and had stolen the full names, social security numbers, birthdates, addresses, and driver's-license numbers of nearly 150 million Americans and 15 million residents of the United Kingdom. Two years later, Equifax agreed to a settlement of $650 million—a sum less than the company's annual cash flow and which provided victims only a few dollars each.

## Multiple National Metaverses

For some 15 years, what we consider "the internet" has become increasingly regionalized. Every country uses the Internet Protocol Suite but the platforms, services, technologies, and conventions in each market have diverged, partly due to the growth in non-American technology giants. Whether it's Europe, Southeast Asia, India, Latin America, China, or Africa, there are more and more successful local start-ups and software leaders than ever before, satisfying everything from payments to groceries and video. If the Metaverse will play an ever-greater role in human culture and labor, then it's also likely its emergence leads to more and stronger regional players.

The most significant cause of fragmentation in the modern internet is nation-specific regulations across the world. The Chinese, European, and Middle Eastern "internets" are increasingly different from those accessed in the United States, Japan, or Brazil due to greater restrictions on data-collection rights, permitted content, and technical standards. As governments around the world contend with the need to regulate the Metaverse—and at the same time, as they try to reduce the power accumulated by the leaders Web 2.0—the world will doubtlessly end up with enormously different outcomes and, dare I say it, "Metaverses."

At the start of this book, I mentioned the South Korean Metaverse Alliance, which was established by the country's Ministry of Science and ICT in mid-2021 and includes more than 450 domestic companies. The organization's specific mandate is not yet clear, but it's likely to be focused on building a stronger Metaverse economy in South Korea, and a larger South Korean presence in the Metaverse globally. To this end, the government will probably drive interoperation and standards that will occasionally disadvantage a given member of the alliance but increase their collective strength, and most importantly benefit South Korea.

Following the trends visible in the Chinese internet today, it's a good bet that China's "Metaverse" will be even more different from (and centrally controlled compared to) that of Western nations. It may arrive much earlier and be more interoperable/standardized, too. Consider Tencent, whose games reach more players, generate more revenue, span more intellectual property, and employ more developers than any other publisher in the world. In China, Tencent releases the titles of companies such as Nintendo, Activision Blizzard, and Square Enix, and develops local editions of hit games such as *PUBG* (which cannot otherwise operate in the country). Tencent's studios are also responsible for the global versions of *Call of Duty Mobile*, *Apex Legends Mobile*, and *PUBG Mobile*. Tencent also owns roughly 40% of Epic Games, 20% of Sea Limited (makers of *Free Fire*), and 15% of Krafton (*PUBG*), and both wholly owns and operates WeChat and QQ, the two most popular messaging apps in China (which also serve as de facto app stores). WeChat is also the second-largest digital payments company/network in China and Tencent already uses facial-recognition software to validate the identity of its players using China's national ID system. No other company is better positioned to facilitate the interoperation of user data, virtual worlds, identity, and payments, nor influence Metaverse standards.

The Metaverse may be a "a massively scaled and interoperable network of real-time rendered 3D virtual worlds," but, as we've seen, it will be realized through physical hardware, computer processors, and networks. Whether those are governed by corporations alone, governments alone, or decentralized groups of tech-savvy coders and developers, the Metaverse is dependent on them. The existence of a virtual tree and its fall may forever be in question, but physics is immutable.

# SPECTATORS, ALL

"TECHNOLOGY FREQUENTLY PRODUCES SURPRISES that no one predicts. But the biggest and most fantastical developments are often anticipated decades in advance." These words opened this book, and in the pages since, hopefully you've come to agree with this observation—and understand its limitations, too. Vannevar Bush had an uncanny ability to predict the devices of the future and much of what they might do, as well as the crucial role of government in making them useful and for the collective benefit. At the same time, his Memex was desk-sized and electromechanical—physically storing and connecting all the content a user might request. Today's pocket-sized, software-operated computers resemble the Memex in spirit alone. In *2001: A Space Odyssey*, Stanley Kubrick imagined a future in which humankind had colonized space and sentient AI had emerged, but iPad-like displays were used for little more than watching TV while eating breakfast and telephones were still dumb and required cords. Neal Stephenson's *Snow Crash* has inspired decades of R&D projects and now guides many of the most powerful companies on earth. Yet Stephenson believed the Metaverse would emerge from the TV industry, not gaming, and was surprised that "instead of people going to bars on the Street in *Snow Crash*, what we have now is *Warcraft* guilds" which go on in-game raids.

I am certain about much of the future. It will be increasingly centered around real-time rendered 3D virtual worlds. Network bandwidth, latency, and reliability will all improve. The amount of computing power will increase, thus enabling higher concurrency, greater persistency, more sophisticated simulations, and altogether new experiences (and yet, the supply of compute will still fall far short of demand for it). Younger generations will be the first to adopt "the Metaverse," and will do so to a greater degree than their parents. Regulators will partly unbundle operating systems, but the companies that own these OSs will still thrive because their unbundled offerings are still market-leading and the emergence of the Metaverse will grow most of these markets. The overall structure of the Metaverse is likely to be similar to those we see today—a handful of horizontally and vertically integrated companies will control a substantial share of the digital economy, with their influence even greater. Regulators will place more scrutiny on them, but will probably still fall short. Some of the major category leaders in the Metaverse will be different from those we know today, while some of today's leaders will be displaced but still survive or even grow. Others will perish. We will continue to use many of the digital and mobile products from the pre-Metaverse era; real-time 3D rendering is not the best way to perform many tasks or experience all forms of content.

Interoperability will be achieved slowly, imperfectly, and never exhaustively or without cost. While the market will eventually solidify around a subset of standards, they won't convert perfectly into one another and each will have drawbacks. And before then, scores of options will be proposed, adopted, deprecated, and forked. Various virtual worlds and integrated virtual world platforms will slowly open up, as was the case with the world economy, while also taking different approaches to the exchange of data and users. For example, many will strike bespoke deals with independent developers, just as the United States has different policies with Canada, Indonesia, Egypt, Honduras, and the European Union (itself a collection of agreements spanning a finite set of "worlds"). There will be taxes, duties, and other fees, as well as the need for multiple identity systems, wallets, and virtual storage lockers. And all policies will be subject to change. The role of blockchain is the least clear aspect of our Metaverse future. To many, it is either critical to the success of the Metaverse or structurally required for it to exist in the first place. Others consider it an interesting technology that will contribute to the Metaverse, but that would exist regardless and in largely the same form. Many consider it an outright scam. Through 2021 and early 2022, blockchains continued to soar, attracting mainstream developers, talented founders, tens of billions in venture capital, even more in institutional investment in cryptocurrencies. And yet, blockchains still have a fairly limited track record of success as of this writing, and the technical, cultural, and legal impediments involved are significant.

By the end of the decade, we'll agree the Metaverse has arrived* and it will be worth many trillions. The question of exactly when it started and how much revenue it generates will remain uncertain. Before getting to that point, we will exit the current phase of hype and probably enter and then exit another one, too. The hype cycle will be caused by at least three factors: the reality that many companies will overpromise what sort of Metaverse experiences will be possible and when; the difficulty of overcoming key technical barriers; and the fact that, even when those barriers are overcome, it will take time to figure out exactly what companies should build "in the Metaverse."

Think back to your first iPhone (or perhaps, your first six). From 2007 to 2013, Apple's operating system was highly skeuomorphic—its iBooks application showed digital versions of books on a digital bookshelf, its notes app was designed to look like a physical yellow pad of paper, its calendar had simulated stitching, and its games center was intended to resemble a felt table. With iOS 7, Apple ditched these legacy design principles for those native to the mobile era. It was during Apple's skeuomorphic era that many of today's leading consumer digital companies were founded. Companies such as Instagram, Snap, and

---

* We may ultimately use a different term for this future due to the extent that the term "Metaverse" is misused, and its potentially negative associations with dystopic science fiction, big tech, blockchains and cryptocurrencies, etc. Recall that in May 2021, Tencent chose to brand its Metaverse efforts "hyper-digital reality," before switching to "Metaverse" as the latter became popular. A reversal of some sort may yet occur.

Slack reimagined what digital communications would be—not using IP to call a landline (Skype) nor text (BlackBerry Messenger), but to reinvent how we communicate, why, and what about. Spotify didn't try to rebroadcast the radio over the internet (Broadcast.com), nor produce internet-only radio (Pandora), but instead changed how we accessed and discovered music. For the foreseeable future, "Metaverse apps" will be stuck in the early stage of development—a videoconference, but in 3D and situated in a simulated corporate boardroom; Netflix, but inside a virtual theater. Slowly, however, we will reinvent everything we do. It is when this process begins, not before, that the Metaverse will feel significant; less like a fantastical vision and more like a practical reality. All of the technologies required to build Facebook were available years before Mark Zuckerberg created the social network. Tinder wasn't invented until five years after the iPhone, at which point 70% of 18-to-34-year-olds had a touchscreen smartphone. Technology is a constraint on the Metaverse, but so is what we imagine and when.

The fits and bursts of Metaverse development will lead to critiques as well as bouts of disappointment and disillusionment. In 1995, Clifford Stoll, an American astronomer and a former systems administer at the US Department of Energy's Lawrence Berkeley National Laboratory, wrote the now-infamous book, *Silicon Snake Oil: Second Thoughts on the Information Highway*. In an editorial for *Newsweek* around the book's publication, he stated that "After two decades online, I'm perplexed . . . uneasy about this most trendy and oversold community. Visionaries see a future of telecommuting workers, interactive libraries and multimedia classrooms. They speak of electronic town meetings and virtual communities. Commerce and business will shift from offices and malls to networks and modems. And the freedom of digital networks will make government more democratic. Baloney. Do our computer pundits lack all common sense . . . what the Internet hucksters won't tell you is that the Internet is one big ocean of unedited data, without any pretense of completeness."[1] Today this reads like a Metaverse criticism that has yet to be published. In December 2000, the *Daily Mail* published a news piece headlined "Internet 'May

Just Be a Passing Fad as Millions Give Up on It,' " backed by research that supposedly estimated Britain was set to lose two million of its 15 million internet users.[2] The critique came after the dotcom crash had begun, at which point the NASDAQ had fallen nearly 40%, but would go on to halve what was left. It took 12 years for the NASDAQ to return to its dotcom-era high. At the time this book went to print, the NASDAQ was more than three times higher than that onetime high.

The future is hard to predict, even for pioneers. We are now on the cusp of the Metaverse, but consider, one last time, the last two eras of computing and networking. Even the most ardent believers in the internet struggled to imagine a future in which there might be billions of web pages across millions of web servers, 300 billion emails per day, with billions of daily users, and a single network, Facebook, counting over three billion monthly users and two billion per day. When he announced the first iPhone in January 2007, Steve Jobs described it as a revolutionary product. He was right, of course. But this first iPhone lacked both an App Store and there were no plans to allow third-party developers to make them. Why? Jobs told developers that "The full Safari engine is inside of iPhone . . . And so, you can write amazing Web 2.0 and Ajax apps that look exactly and behave exactly like apps on the iPhone."[3] But in October 2007, ten months after the iPhone had been unveiled and four months after it went on sale, Jobs changed his mind. An SDK was announced for March 2008, with the App Store released in July of that year. Within a month, the million or so iPhone owners had downloaded 30% as many apps as the more than 40 million iTunes users had downloaded songs. Jobs then told the *Wall Street Journal*: "I would not trust any of our predictions because reality has so far exceeded them by such a great degree that we've been reduced to spectators just like you, watching this amazing phenomenon."[4]

The trajectory of the Metaverse will be broadly similar. Whenever a technological breakthrough occurs, consumers, developers, and entrepreneurs respond. Eventually, a thing that seems trivial—a mobile phone, a touchscreen, a video game—becomes essential, and ends up changing the world in ways both predicted and never even considered.

# ACKNOWLEDGMENTS

This book exists thanks to the many family members, advocates, teachers, friends, entrepreneurs, dreamers, writers, and creators who have inspired and taught me over the past four decades. Here is just a small selection of these individuals. Jo-Anne Boluk, Ted Ball, Poppo, Brenda and Al Harrow, Anshul Ruparell, Michael Zawalsky, Will Meneray, Abhinav Saksena, Jason Hirschhorn, Chris Meledandri, Tal Shachar, Jack Davis, Julie Young, Gady Epstein, Jacob Navok, Chris Cataldi, Jayson Chi, Sophia Feng, Anna Sweet, Imran Sarwar, Jonathan Glick, Peter Rojas, Peter Kafka, Matthew Henick, Sharon Tal Yguado, Kuni Takahashi, Tony Driscoll, Mark Noseworthy, Amanda Moon, Thomas LeBien, Daniel Geistle, Pilar Queen, Charlotte Perman, Paul Rehrig, and Gregory McDonald.

# NOTES

### Introduction

1. Casey Newton, "Mark in the Metaverse: Facebook's CEO on Why the Social Network Is Becoming 'a Metaverse Company,' " *The Verge*, July 22, 2021, accessed January 4, 2022, https://www.theverge.com/.
2. Dean Takahashi, "Nvidia CEO Jensen Huang Weighs in on the Metaverse, Blockchain, and Chip Shortage," *Venture Beat*, June 12, 2021, accessed January 4, 2022, https://venturebeat.com/.
3. Data pulled from Bloomberg database on January 2, 2022 (excludes a dozen references to companies that included "Metaverse" only in their names).
4. Zheping Huang, "Tencent Doubles Social Aid to $15 Billion as Scrutiny Grows," *Bloomberg*, August 18, 2021, accessed January 4, 2022, https://www.bloomberg.com/.
5. Chang Che, "Chinese Investors Pile into 'Metaverse,' Despite Official Warnings," *SupChina*, September 24, 2021, accessed January 4, 2021, https://supchina.com/2021/09/24/chinese-investors-pile-into-metaverse-despite-official-warnings/.
6. Jens Bostrup, "EU's Danske Chefforhandler: Facebooks store nye projekt 'Metaverse' er dybt bekymrende," *Politiken*, October 18, 2021, accessed January 4, 2022, https://politiken.dk/.

### Chapter 1    A Brief History of the Future

1. Neal Stephenson, *Snow Crash* (New York: Random House, 1992), 7.
2. John Schwartz, "Out of a Writer's Imagination Came an Interactive World," *New York Times*, December 5, 2011, accessed January 4, 2022, https://www.nytimes.com/.
3. Joanna Robinson, "The Sci-Fi Guru Who Predicted Google Earth Explains Silicon Valley's Latest Obsession," *Vanity Fair*, June 23, 2017, accessed January 4, 2022, https://www.vanityfair.com/.
4. Stanley Grauman Weinbaum, *Pygmalion's Spectacles* (1935), Kindle edition, p. 2.

5. Ryan Zickgraf, "Mark Zuckerberg's 'Metaverse' Is a Dystopian Nightmare," *Jacobin*, September 25, 2021, accessed January 4, 2022, https://www.jacobinmag.com/.

6. J. D. N. Dionisio, W. G. Burns III, and R. Gilbert, "3D Virtual Worlds and the Metaverse: Current Status and Future Possibilities," *ACM Computing Surveys* 45, issue 3 (June 2013), http://dx.doi.org/10.1145/2480741.2480751.

7. Josh Ye, "One Gamer Spent a Year Building This Cyberpunk City in Minecraft," *South China Morning Post,* January 15, 2019, accessed January 4, 2022, https://www.scmp.com/.

8. Josh Ye, "Minecraft Players Are Recreating China's Rapidly Built Wuhan Hospitals," *South China Morning Post,* February 20, 2020, accessed January 4, 2022, https://www.scmp.com/.

9. Tim Sweeney (@TimSweeneyEpic), Twitter, June 13, 2021, accessed January 4, 2022, https://twitter.com/timsweeneyepic/status/1404241848147775488.

10. Tim Sweeney (@TimSweeneyEpic), Twitter, June 13, 2021, accessed January 4, 2022, https://twitter.com/TimSweeneyEpic/status/1404242449053241345?s=20.

11. Dean Takahashi, "The DeanBeat: Epic Graphics Guru Tim Sweeney Foretells How We Can Create the Open Metaverse," *Venture Beat*, December 9, 2016, accessed January 4, 2022, https://venturebeat.com/.

## Chapter 2  Confusion and Uncertainty

1. Satya Nadella, "Building the Platform for Platform Creators," LinkedIn, May 25, 2021, accessed January 4, 2022 https://www.linkedin.com/pulse/building-platform-creators-satya-nadella.

2. Sam George, "Converging the Physical and Digital with Digital Twins, Mixed Reality, and Metaverse Apps," Microsoft Azure, May 26, 2021, accessed January 4, 2022, https://azure.microsoft.com/en-ca/blog/converging-the-physical-and-digital-with-digital-twins-mixed-reality-and-metaverse-apps/.

3. Andy Chalk, "Microsoft Says It Has Metaverse Plans for Halo, Minecraft, and Other Games," *PC Gamer*, November 2, 2021, accessed January 4, 2022, https://www.pcgamer.com/microsoft-says-it-has-metaverse-plans-for-halo-minecraft-and-other-games/.

4. Gene Park, "Epic Games Believes the Internet Is Broken. This Is Their Blueprint to Fix It," *Washington Post*, September 28, 2021, accessed January 4, 2022, https://www.washingtonpost.com/video-games/2021/09/28/epic-fortnite-metaverse-facebook/.

5. Alex Sherman, "Execs Seemed Confused About the Metaverse on Q3 Earnings Calls," CNBC, November 20, 2021, accessed January 5, 2021, https://www.cnbc.com/2021/11/20/executives-wax-poetic-on-the-metaverse-during-q3-earnings-calls.html.

6. CNBC, "Jim Cramer Explains the 'Metaverse' and What It Means for Facebook," July 29, 2021, accessed January 5, 2022, https://www.cnbc.com/video/2021/07/29/jim-cramer-explains-the-metaverse-and-what-it-means-for-facebook.html.

7. Elizabeth Dwoskin, Cat Zakrzewski, and Nick Miroff, "How Facebook's 'Metaverse' Became a Political Strategy in Washington," *Washington Post*, Sep-

tember 24, 2021, accessed January 3, 2022, https://www.washingtonpost.com/technology/2021/09/24/facebook-washington-strategy-metaverse/.

8.  Tim Sweeney (@TimSweeneyEpic), Twitter, August 6, 2020, accessed January 4, 2022, https://twitter.com/timsweeneyepic/status/1291509151567425536.

9.  Alaina Lancaster, "Judge Gonzalez Rogers Is Concerned That Epic Is Asking to Pay Apple Nothing," *The Law*, May 24, 2021, accessed June 2, 2021, https://www.law.com/therecorder/2021/05/24/judge-gonzalez-rogers-is-concerned-that-epic-is-asking-to-pay-apple-nothing/?slreturn=20220006091008.

10. John Koetsier, "The 36 Most Interesting Findings in the Groundbreaking Epic Vs Apple Ruling That Will Free The App Store," *Forbes*, September 10, 2021, accessed January 3, 2022, https://www.forbes.com/sites/johnkoetsier/2021/09/10/the-36-most-interesting-findings-in-the-groundbreaking-epic-vs-apple-ruling-that-will-free-the-app-store/?sh=56db5566fb3f.

11. Wikipedia, s.v. "Internet," last edited October 13, 2021, https://en.wikipedia.org/wiki/Internet.

12. Paul Krugman, "Why Most Economists' Predictions Are Wrong," *Red Herring Online*, June 10, 1998, Internet Archive, https://web.archive.org/web/19980610100009/http://www.redherring.com/mag/issue55/economics.html.

13. Wired Staff, "May 26, 1995: Gates, Microsoft Jump on 'Internet Tidal Wave,' " *Wired*, May 26, 2021, accessed January 5, 2022, https://www.wired.com/2010/05/0526bill-gates-internet-memo/.

14. CNBC, "Microsoft's Ballmer Not Impressed with Apple iPhone," January 17, 2007, accessed January 4, 2022, https://www.cnbc.com/id/16671712.

15. Drew Olanoff, "Mark Zuckerberg: Our Biggest Mistake Was Betting Too Much On HTML5," *TechCrunch*, September 11, 2022, accessed January 5, 2022, https://techcrunch.com/2012/09/11/mark-zuckerberg-our-biggest-mistake-with-mobile-was-betting-too-much-on-html5/.

16. M. Mitchell Waldrop, *Complexity: The Emerging Science at the Edge of Order and Chaos* (New York: Simon & Schuster, 1992), 155.

### Chapter 3    A Definition (Finally)

1.  Dean Takahashi, "How Pixar Made Monsters University, Its Latest Technological Marvel," *Venture Beat*, April 24, 2013, accessed January 5, 2022, https://venturebeat.com/2013/04/24/the-making-of-pixars-latest-technological-marvel-monsters-university/.

2.  Wikipedia, s.v. "Metaphysics," last edited October 28, 2021, https://en.wikipedia.org/wiki/Metaphysics.

3.  Stephenson, *Snow Crash*, 27.

4.  CCP Team, "Infinite Space: An Argument for Single-Sharded Architecture in MMOs," *Game Developer*, August 9, 2010, accessed January 5, 2022, https://www.gamedeveloper.com/design/infinite-space-an-argument-for-single-sharded-architecture-in-mmos.

5.  "John Carmack Facebook Connect 2021 Keynote," posted by Upload VR, October 28, 2021, accessed January 5, 2022, https://www.youtube.com/watch?v=BnSUk0je6oo.

### Chapter 4    The Next Internet

1.  Josh Stark and Evan Van Ness, "The Year in Ethereum 2021," *Mirror*, January 17, 2022, accessed February 2, 2022, https://stark.mirror.xyz/q3OnsK7mvfGt TQ72nfoxLyEV5lfYOqUfJIoKBx7BG1I.
2.  BBC, "Military Fears over PlayStation2," April 17, 2000, accessed January 4, 2022, http://news.bbc.co.uk/2/hi/asia-pacific/716237.stm.
3.  "Secretary of Commerce Don Evans Applauds Senate Passage of Export Administration Act as Modern-day Legislation for Modern-day Technology," Bureau of Industry and Security, US Department of Commerce, 6 September 2001, www.bis.doc.gov.
4.  Chas Littell, "AFRL to Hold Ribbon Cutting for Condor Supercomputer," Wright-Patterson Air Force Base, press release, November 17, 2010, accessed January 5, 2022, https://www.wpafb.af.mil/News/Article-Display/ Article/399987/afrl-to-hold-ribbon-cutting-for-condor-supercomputer/.
5.  Lisa Zyga, "US Air Force Connects 1,760 PlayStation 3's to Build Supercomputer," Phys.org, December 2, 2010, accessed January 5, 2022, https://phys .org/news/2010-12-air-playstation-3s-supercomputer.html.
6.  Even Shapiro, "The Metaverse Is Coming. Nvidia CEO Jensen Huang on the Fusion of Virtual and Physical Worlds," *Time*, April 18, 2021, accessed January 2, 2022, https://time.com/5955412/artificial-intelligence-nvidia-jensen -huang/.
7.  David M. Ewalt, "Neal Stephenson Talks About Video Games, the Metaverse, and His New Book, REAMDE," *Forbes,* September 19, 2011.
8.  Daniel Ek, "Daniel Ek—Enabling Creators Everywhere," *Colossus,* September 14, 2021, accessed January 5, 2022, https://www.joincolossus.com/ episodes/14058936/ek-enabling-creators-everywhere?tab=transcript.
9.  David M. Ewalt, "Neal Stephenson Talks About Video Games, the Metaverse, and His New Book, REAMDE," *Forbes,* September 19, 2011.

### Chapter 5    Networking

1.  Farhad Manjoo, "I Tried Microsoft's Flight Simulator. The Earth Never Seemed So Real," *New York Times*, August 19, 2022, accessed January 4, 2022, https://www.nytimes.com/2020/08/19/opinion/microsoft-flight-simulator .html.
2.  Seth Schiesel, "Why Microsoft's New Flight Simulator Should Make Google and Amazon Nervous," *Protocol*, August 16, 2020, accessed January 5, 2022, https://www.protocol.com/microsoft-flight-simulator-2020.
3.  Eryk Banatt, Stefan Uddenberg, and Brian Scholl, "Input Latency Detection in Expert-Level Gamers," Yale University, April 21, 2017, accessed January 4, 2022, https://cogsci.yale.edu/sites/default/files/files/Thesis2017Banatt.pdf.
4.  Rob Pegoraro, "Elon Musk: 'I Hope I'm Not Dead by the Time People Go to Mars,' " *Fast Company,* March 10, 2020, accessed January 3, 2022, https:// www.fastcompany.com/90475309/elon-musk-i-hope-im-not-dead-by-the -time-people-go-to-mars.

### Chapter 6    Computing

1.  Foundry Trends, "One Billion Assets: How Pixar's Lightspeed Team Tackled Coco's Complexity," October 25, 2018, accessed January 5, 2022, https://www .foundry.com/insights/film-tv/pixar-tackled-coco-complexity.
2.  Dean Takahashi, "Nvidia CEO Jensen Huang Weighs in on the Metaverse, Blockchain, and Chip Shortage," *Venture Beat*, June 12, 2021, accessed February 1, 2022, https://venturebeat.com/2021/06/12/nvidia-ceo-jensen-huang-weighs -in-on-the-metaverse-blockchain-chip-shortage-arm-deal-and-competition/.
3.  Raja Koduri, "Powering the Metaverse," Intel, December 14, 2021, accessed January 4, 2022, https://www.intel.com/content/www/us/en/newsroom/opinion/ powering-metaverse.html.
4.  Tim Sweeney (@TimSweeneyEpic), Twitter, January 7, 2020, accessed January 4, 2022, https://twitter.com/timsweeneyepic/status/1214643203871248385.
5.  Peter Rubin, "It's a Short Hop from Fortnite to a New AI Best Friend," *Wired*, March 21, 2019, accessed February 1, 2021, https://www.wired.com/story/ epic-games-qa/.

### Chapter 7    Virtual World Engines

1.  " 'The Future—It's Bigger and Weirder than You Think—' by Owen Mahoney, NEXON CEO," posted by NEXON, December 20, 2019, accessed January 5, 2022, https://www.youtube.com/watch?v=VqiwZN1CShI.
2.  Roblox, "A Year on Roblox: 2021 in Data," January 26, 2022, accessed February 3, 2022, https://blog.roblox.com/2022/01/year-roblox-2021-data/.

### Chapter 8    Interoperability

1.  Josh Ye (@TheRealJoshYe), Twitter, May 3, 2021, accessed February 1, 2022, https://mobile.twitter.com/therealjoshye/status/1389217569228296201.
2.  Tom Phillips, "So, Will Sony Actually Allow PS4 and Xbox One Owners to Play Together?," *Eurogamer*, March 17, 2016, accessed January 5, 2022, https:// www.eurogamer.net/articles/2016-03-17-sonys-shuhei-yoshida-on-playstation -4-and-xbox-one-cross-network-play.
3.  Jay Peters, "Fortnite's Cash Cow Is PlayStation, Not iOS, Court Documents Reveal," *The Verge*, April 28, 2021, accessed February 1, 2022, https:// www.theverge.com/2021/4/28/22407939/fortnite-biggest-platform-revenue -playstation-not-ios-iphone.
4.  Aaron Rakers, Joe Quatrochi, Jake Wilhelm, and Michael Tsevtanov, "NVDA: Omniverse Enterprise—Appreciating NVIDIA's Platform Strategy to Capitalize ($10B+) on the 'Metaverse,'" *Wells Fargo*, November 3, 2021.
5.  Chris Michaud, "English the Preferred Language for World Business: Poll," Reuters, May 12, 2016, https://www.reuters.com/article/us-language/english- the-preferred-language-for-world-business-poll-idUSBRE84F0OK20120516.
6.  Epic Games, "Tonic Games Group, Makers of 'Fall Guys', Joins Epic Games," March 2, 2021, accessed February 2, 2022, https://www.epicgames.com/site/ en-US/news/tonic-games-group-makers-of-fall-guys-joins-epic-games.

### Chapter 9    Hardware

1.  Mark Zuckerberg, Facebook, April 29, 2021, accessed January 5, 2022, https://www.facebook.com/zuck/posts/the-hardest-technology-challenge-of-our-time-may-be-fitting-a-supercomputer-into/10112933648910701/.
2.  Tech@Facebook, "Imagining a New Interface: Hands-Free Communication without Saying a Word," March 30, 2020, accessed January 4, 2022, https://tech.fb.com/imagining-a-new-interface-hands-free-communication-without-saying-a-word/.
3.  Tech@Facebook, "BCI Milestone: New Research from UCSF with Support from Facebook Shows the Potential of Brain-Computer Interfaces for Restoring Speech Communication," July 14, 2021, accessed January 4, 2022, https://tech.fb.com/bci-milestone-new-research-from-ucsf-with-support-from-facebook-shows-the-potential-of-brain-computer-interfaces-for-restoring-speech-communication/.
4.  Antonio Regalado, "Facebook Is Ditching Plans to Make an Interface that Reads the Brain," *MIT Technology Review*, July 14, 2021, accessed January 4, 2022, https://www.technologyreview.com/2021/07/14/1028447/facebook-brain-reading-interface-stops-funding/.
5.  Andrew Nartker, "How We're Testing Project Starline at Google," Google Blog, November 30, 2021, accessed February 2, 2022, https://blog.google/technology/research/how-were-testing-project-starline-google/.
6.  Will Marshall, "Indexing the Earth," *Colossus*, November 15, 2021, accessed January 5, 2022, https://www.joincolossus.com/episodes/14029498/marshall-indexing-the-earth?tab=blocks.
7.  Nick Wingfield, "Unity CEO Predicts AR-VR Headsets Will Be as Common as Game Consoles by 2030," *The Information*, June 21, 2021.


### Chapter 10    Payment Rails

1.  NACHA, "ACH Network Volume Rises 11.2% in First Quarter as Two Records Are Set," press release, April 15, 2021, accessed January 26, 2022, https://www.prnewswire.com/news-releases/ach-network-volume-rises-11-2-in-first-quarter-as-two-records-are-set-301269456.html.
2.  Takashi Mochizuki and Vlad Savov, "Epic's Battle with Apple and Google Actually Dates Back to Pac-Man," *Bloomberg*, August 19, 2020, accessed January 4, 2021, https://www.bloomberg.com/.
3.  Tim Sweeney (@TimSweeneyEpic), Twitter, January 11, 2020, accessed January 4, 2022, https://twitter.com/TimSweeneyEpic/status/1216089159946948620.
4.  Epic Games, "Epic Games Store Weekly Free Games in 2020!," January 14, 2022, accessed February 14, 2022, https://www.epicgames.com/store/en-US/news/epic-games-store-weekly-free-games-in-2020.
5.  Epic Games, "Epic Games Store 2020 Year in Review," January 28, 2021, accessed February 14, 2022, https://www.epicgames.com/store/en-US/news/epic-games-store-2020-year-in-review.
6.  Epic Games, "Epic Games Store 2021 Year in Review," January 27, 2022, accessed February 14, 2022, https://www.epicgames.com/store/en-US/news/epic-games-store-2021-year-in-review.

7. Tyler Wilde, "Epic Will Lose Over $300M on Epic Games Store Exclusives, Is Fine With That," *PC Gamer*, April 10, 2021, accessed February 14, 2022, https://www.pcgamer.com/epic-games-store-exclusives-apple-lawsuit/.

8. Adi Robertson, "Tim Cook Faces Harsh Questions about the App Store from Judge in Fortnite Trial," *The Verge*, May 21, 2021, accessed January 5, 2022, https://www.theverge.com/2021/5/21/22448023/epic-apple-fortnite-antitrust-lawsuit-judge-tim-cook-app-store-questions.

9. Nick Wingfield, "IPhone Software Sales Take Off: Apple's Jobs," *Wall Street Journal*, August 11, 2008.

10. John Gruber, "Google Announces Chrome for iPhone and iPad, Available Today," *Daring Fireball*, June 28, 2021, accessed January 4, 2022, https://daringfireball.net/linked/2012/06/28/chrome-ios.

11. Kate Rooney, "Apple: Don't Use Your iPhone to Mine Cryptocurrencies," *CNBC*, June 11, 2018, accessed January 4, 2021, https://www.cnbc.com/2018/06/11/dont-even-think-about-trying-to-bitcoin-with-your-iphone.html.

12. Tim Sweeney (@TimSweeneyEpic), Twitter, February 4, 2022, accessed February 5, 2022, https://twitter.com/TimSweeneyEpic/status/1489690359194173450.

13. Marco Arment (@MarcoArment), Twitter, February 4, 2022, accessed February 5, 2022, https://twitter.com/marcoarment/status/1489599440667168768.

14. Manoj Balasubramanian, "App Tracking Transparency Opt-In Rate—Monthly Updates," *Flurry*, December 15, 2021, accessed February 5, 2022, https://www.flurry.com/blog/att-opt-in-rate-monthly-updates/.

### *Chapter 11*    **Blockchains**

1. Telegraph Reporters, "What Is Ethereum and How Does It Differ from Bitcoin?," *The Telegraph*, August 17, 2018.

2. Ben Gilbert, "Almost No One Knows about the Best Android Phones on the Planet," *Insider*, October 25, 2015, accessed January 4, 2022, https://www.businessinsider.com/why-google-makes-nexus-phones-2015-10.

3. Wikipedia, s.v. "Possession is Nine-Tenths of the Law," last edited December 6, 2021, https://en.wikipedia.org/wiki/Possession_is_nine-tenths_of_the_law.

4. Hannah Murphy and Joshua Oliver, "How NFTs Became a $40bn Market in 2021," *Financial Times*, December 31, 2021, accessed January 4, 2022. Note, this sum, $40.9 billion, is limited to the Ethereum blockchain, which is estimated to have 90% share of NFT transactions.

5. Kevin Roose, "Maybe There's a Use for Crypto After All," *New York Times*, February 6, 2022, accessed February 7, 2022, https://www.nytimes.com/2022/02/06/technology/helium-cryptocurrency-uses.html.

6. Kevin Roose, "Maybe There's a Use for Crypto After All," *New York Times*, February 6, 2022, accessed February 7, 2022, https://www.nytimes.com/2022/02/06/technology/helium-cryptocurrency-uses.html.

7. Helium, accessed March 5, 2022, https://explorer.helium.com/hotspots.

8. CoinMarketCap, "Helium," accessed February 7, 2022, https://coinmarketcap.com/currencies/helium/.

9. Dean Takahashi, "The DeanBeat: Predictions for gaming in 2022," *Venture Beat*, December 31, 2021, accessed January 3, 2022, https://venturebeat.com/2021/12/31/the-deanbeat-predictions-for-gaming-2022/.

10. Ephrat Livni, "Venture Capital Funding for Crypto Companies Is Surging," *New York Times*, December 1, 2021, accessed January 5, 2022, https://www
.nytimes.com/2021/12/01/business/dealbook/crypto-venture-capital.html.

11. Olga Kharif, "Crypto Crowdfunding Goes Mainstream with Constitu-tionDAO Bid," *Bloomberg*, November 20, 2021, accessed January 2, 2022, https://www.bloomberg.com/news/articles/2021-11-20/crypto-crowdfund-ing-goes-mainstream-with-constitutiondao-bid?sref=sWz3GEG0.

12. Miles Kruppa, "Crypto Assets Inspire New Brand of Collectivism Beyond Finance," *Financial Times*, December 27, 2021, accessed January 4, 2022, https://www.ft.com/content/c4b6d38d-e6c8-491f-b70c-7b5cf8f0cea6.

13. Lizzy Gurdus, "Nvidia CEO Jensen Huang: Cryptocurrency Is Here to Stay, Will Be an 'Important Driver' For Our Business," CNBC, March 29, 2018, accessed February 2, 2022, https://www.cnbc.com/2018/03/29/nvidia-ceo-jensen-huang-cryptocurrency-blockchain-are-here-to-stay.html.

14. Visa, "Crypto: Money Is Evolving," accessed February 2, 2022, https://usa.visa
.com/solutions/crypto.html.

15. Dean Takahashi, "Game Boss Interview: Epic's Tim Sweeney on Blockchain, Digital Humans, and Fortnite," *Venture Beat*, August 30, 2017, accessed February 2, 2022, https://venturebeat.com/2017/08/30/game-boss-interview-epics-tim-sweeney-on-blockchain-digital-humans-and-fortnite/.

16. Tim Sweeney (@TimSweeneyEpic), Twitter, January 30, 2021, accessed January 4, 2022, https://twitter.com/TimSweeneyEpic/status/1355573241964802050.

17. Tim Sweeney (@TimSweeneyEpic), Twitter, September 27, 2021, accessed January 4, 2022, https://twitter.com/TimSweeneyEpic/status/1442519522875949061.

18. Tim Sweeney (@TimSweeneyEpic), Twitter, October 15, 2021, accessed January 4, 2022, https://twitter.com/TimSweeneyEpic/status/1449146317129895938.

### Chapter 12    When Will the Metaverse Arrive?

1. Tom Huddleston Jr., "Bill Gates Says the Metaverse Will Host Most of Your Office Meetings Within 'Two or Three Years'—Here's What It Will Look Like," CNBC, December 9, 2021, accessed February 2, 2022, https://www
.cnbc.com/2021/12/09/bill-gates-metaverse-will-host-most-virtual-meetings-in-a-few-years.html.

2. "The Metaverse and How We'll Build It Together—Connect 2021," posted by Meta, October 28, 2021, accessed February 2, 2022, https://www.youtube
.com/watch?v=Uvufun6xer8.

3. Steven Ma, "Video Games' Future Is More Than the Metaverse: Let's Talk 'Hyper Digital Reality'," *GamesIndustry*, February 8, 2022, accessed February 11, 2022, https://www.gamesindustry.biz/articles/2022-02-07-the-future-of-games-is-far-more-than-the-metaverse-lets-talk-hyper-digital-reality.

4. George Smiley, "The U.S. Economy in the 1920s," Economic History Association, accessed January 5, 2022, https://eh.net/encyclopedia/the-u-s-economy-in-the-1920s/.

5. Tim Hartford, "Why Didn't Electricity Immediately Change Manufactur-ing?," August 21, 2017, accessed January 5, 2022, https://www.bbc.com/news/business-40673694.

6. David E. Nye, *America's Assembly Line* (Cambridge, MA: MIT Press, 2015), 19.

## Chapter 13    Meta-Businesses

1.  Wikipedia, s.v. "Baumol's cost disease," last edited October 2, 2022, https://en.wikipedia.org/wiki/Baumol%27s_cost_disease.
2.  US Bureau of Labor Statistics, accessed December 2021.
3.  Melissa Pankida, "The Psychology Behind Why We Speed Swipe on Dating Apps," *Mic*, September 27, 2019, accessed January 2, 2022, https://www.mic.com/life/we-speed-swipe-on-tinder-for-different-reasons-depending-on-our-gender-18808262.
4.  Benedict Evans, "Cars, Newspapers and Permissionless Innovation," September 6, 2015, accessed January 2, 2022, https://www.ben-evans.com/benedictevans/2015/9/1/permissionless-innovation.
5.  Gene Park, "Epic Games Believes the Internet Is Broken. This Is Their Blueprint to Fix It," *Washington Post*, September 28, 2021, accessed January 4, 2022, https://www.washingtonpost.com/video-games/2021/09/28/epic-fortnite-metaverse-facebook/.
6.  Bob Woods, "The First Metaverse Experiments? Look to What's Already Happening in Medicine," CNBC, December 4, 2021, accessed January 4, 2022, https://www.cnbc.com/2021/12/04/the-first-metaverse-experiments-look-to-whats-happening-in-medicine.html.

## Chapter 14    Metaverse Winners and Losers

1.  Microsoft, "Microsoft to Acquire Activision Blizzard to Bring the Joy and Community of Gaming to Everyone, Across Every Device," January 18, 2022, accessed February 2, 2022, https://news.microsoft.com/2022/01/18/microsoft-to-acquire-activision-blizzard-to-bring-the-joy-and-community-of-gaming-to-everyone-across-every-device/.
2.  Adi Robertson, "Tim Cook Faces Harsh Questions about the App Store from Judge in Fortnite Trial," *The Verge*, May 21, 2021, accessed January 4, 2022, https://www.theverge.com/2021/5/21/22448023/epic-apple-fortnite-antitrust-lawsuit-judge-tim-cook-app-store-questions.
3.  Brad Smith, "Adapting Ahead of Regulation: A Principled Approach to App Stores," Microsoft, February 9, 2022, accessed February 11, 2022, https://blogs.microsoft.com/on-the-issues/2022/02/09/open-app-store-principles-activision-blizzard/
4.  Brad Smith (@BradSmi), Twitter, February 3, 2022, accessed February 4, 2022, https://twitter.com/BradSmi/status/1489395484808466438.

## Chapter 15    Metaversal Existence

1.  Sean Hollister, "Here's What Apple's New Rules about Cloud Gaming Actually Mean," *The Verge*, September 18, 2020, accessed January 4, 2022, https://www.theverge.com/2020/9/18/20912689/apple-cloud-gaming-streaming-xcloud-stadia-app-store-guidelines-rules.

## *Conclusion* Spectators, All

1. Clifford Stoll, "Why the Web Won't Be Nirvana," *Newsweek*, February 26, 1995, accessed January 6, 2022, https://www.newsweek.com/clifford-stoll-why-web-wont-be-nirvana-185306.

2. James Chapman, "Internet 'May Just Be a Passing Fad as Millions Give Up on It,' " *Daily Mail*, December 5, 2000.

3. 9to5 Staff, "Jobs' Original Vision for the iPhone: No Third-Party Native Apps," *9to5Mac,* October 21, 2011, accessed January 5, 2022, https://9to5mac .com/2011/10/21/jobs-original-vision-for-the-iphone-no-third-party-native -apps/.

4. Nick Wingfield, " 'The Mobile Industry's Never Seen Anything Like This': An Interview with Steve Jobs at the App Store's Launch," *Wall Street Journal,* originally recorded August 7, 2008, published in full on July 25, 2018, accessed January 5, 2022, https://www.wsj.com/articles/the-mobile-industrys-never-seen -anything-like-this-an-interview-with-steve-jobs-at-the-app-stores-launch -1532527201.

# INDEX

Page numbers in *italics* refer to illustrations and tables.